The Finances of Engineering Companies
An introduction for students and practising engineers

T0221298

The Finances of Engineering Companies

An introduction for Students and Practising Engineers

A J Reynolds

Pro-Vice-Chancellor and Professor of Mechanical
Engineering, Brunel University

Routledge
Taylor & Francis Group

LONDON AND NEW YORK

First published in 1992 by Edward Arnold.
Subsequently published by Butterworth-Heinemann

This edition published 2011 by Routledge
2 Park Square, Milton Park, Abingdon, Oxon OX14 4RN
711 Third Avenue, New York, NY 10017, USA

Routledge is an imprint of the Taylor & Francis Group, an informa business

First published 1992

British Library Cataloguing in Publication Data
A catalogue record for this book is available from the British
Library

Library of Congress Cataloguing in Publication Data
A catalogue record for this book is available from the Library
of Congress

ISBN 978-0-415-50323-5

Contents

Preface

This book presents a course in company finance for students of engineering, probably in the final year of a degree course. It will also be useful to qualified engineers who seek insight into the business of engineering. Initially, the course met the needs of students of mechanical engineering, but the requirements of other fields are rather similar, and the course is now taken by students of both mechanical and electrical engineering. The broadening of the range of participants has led to the introduction of a wide range of illustrative examples, which is to the advantage of all concerned.

When preparing lectures to introduce students to this aspect of business, the author was not able to find textbooks that met fully his needs and theirs. The available treatments are written for students preparing for careers in business or finance, and do not try to relate general principles to the affairs of the companies of particular interest to engineers. Nor do these texts provide the overview that is appropriate in the context of a curriculum which makes many other demands on students. In content and approach this book seeks to address the limitations of existing introductions to company finance.

Many of the illustrative examples relate to British companies prominent in fields that can readily be identified as engineering – such as British Aerospace, General Electric, Trafalgar House, British Steel and Imperial Chemical Industries. However, to put the affairs of these companies in perspective, consideration is given to more varied commercial endeavours – Glaxo Holdings, J Sainsbury and the Ladbroke Group – and to overseas companies – the American Ford Motor Company and Volvo, the diversified Swedish company best known for its cars and commercial vehicles.

Most of the cases chosen to illustrate principles are topical, relating to events only a year or so before the publication of this book. While equally relevant examples will doubtless emerge in the years to come, many of those selected will retain their relevance and utility. That said, changes in reporting practice – and some significant changes are in train – will necessitate alterations in details within a few years. A number of changes known to be imminent have been foreshadowed.

A further challenge in preparing a book of this kind is the pace of change in corporate organisation. At the time of writing, the future structure and ownership of three prominent manufacturing companies are under active debate. It has been suggested that British Aerospace, GEC and ICI should be broken down into smaller, more focused components. Nothing may come of these proposals, but they illustrate the changes in ownership and organisation that can affect even the largest companies. This is a tantalising subject on which to write, for both the players and the rules of the game are subject to continual change.

As is always the case, I have learned much from the response of real students, rather than those imagined when the course was being developed. Some of the examples included in this book were created to resolve difficulties that my students had with concepts – even modes of thinking – unlike those required in other aspects of their studies. The preparation of this book has caused me to think further about many aspects of finance, and to sharpen the discussion in a number of ways. Accordingly, I have to offer both thanks and apologies to the students who have influenced my thinking and the form of this book.

A J Reynolds
1992

1

What is this Book About?

This book introduces student engineers to a crucially important aspect of engineering. It is concerned with what companies are, how their performance can be described and measured, and why they succeed or fail financially. The most innovative research, the most ingenious products, and the most modern production lines will not bring success if the finances of the company are poorly organised.

Some features of the business of engineering, such as costing and the management of specific projects, have long had a place in the curriculum. Suitable textbooks are available to define and support courses in those areas. But when engineering teachers and students turn to another important aspect of their business – the overall finances of engineering businesses and the holding companies that control them – they are unlikely to find teaching materials that meet their needs.

What are the special requirements of engineering students? They need a compact and focused treatment of business finance, for there are many other calls on their time. That treatment can assume little prior knowledge of the world of business and its special vocabulary. It must equip engineers to understand and interpret financial statements, but need not prepare them to generate accounts without support from professional accountants or extensive further training. Finally, engineering students are better able to see the relevance of company finance if reference is made to organisations they know, and in which they may work at some time in their careers.

1.1 The Role of the Company

In a market-driven economy the organisations responsible for the generation of most physical products and many kinds of service are companies with limited liability. There are alternative ways of organising economic activity. The planned economies of the Communist bloc have operated quite differently, largely organised centrally by agencies of the State. Even within nominally capitalist economies, important services – education usually, and medical provision sometimes – are provided by the State or by state-controlled organisations. Business structures such as partnerships can operate stably, though generally on a much smaller scale, in economies dominated by much larger corporations.

The Public Limited Company

As the word implies, a *Company* is a grouping of like-minded individuals, who create or sustain the organisation for a common purpose or purposes. The label *Limited Liability* is of crucial importance; it indicates that the financial obligations of the owners or *Members* of the company are predetermined. The limited company is a distinct 'legal person', quite separate from its members. Hence they are not exposed to unlimited loss if the company of which they are members incurs large financial liabilities. All they can lose is the amount they have subscribed to join the company. This leaves unquestioned the extent of the moral responsibility of the 'owners' or 'members' for the activities of the company and its employees.

Participation in the ownership of a company is usually defined by the ownership of *Shares* issued by the company. The owners, or *Shareholders*, usually participate in profits generated by the company's operations in proportion to their shareholdings, and exercise control over it in the same proportion. However, a company may issue distinct classes of shares, entitling their owners to different benefits and levels of control over its affairs.

The management of a company is entrusted to *Directors*, elected by the shareholders to exercise supervision on their behalf, subject to periodic review, usually annually. Collectively, they are referred to as the *Board of Directors*, or simply *The Board*. Directors are usually shareholders themselves, and sometimes own significant numbers of shares. Day-to-day operations are carried out by managers and employees, who may also be shareholders or directors.

As an example, consider British Aerospace plc, one of the largest UK-registered engineering companies. At the end of 1990 its Board consisted of fourteen directors, of whom four were described as non-executive, that is, independent persons who might be expected to represent the interests of the shareholders. The company had nearly 92,000 shareholders, and the group of companies it controlled employed some 128,000 people.

The form of organisation that is dominant in the United Kingdom and similar economies is the *Public Limited Company*. In the United Kingdom this is defined as a company whose share capital exceeds a specified value (currently £50,000) and which has allotted to shareholders at least one-quarter of its shares. Only a Public Limited Company (abbreviated to PLC, Plc or plc, or to ccc in Wales) can have its shares quoted and traded on a stock exchange. Equivalent organisations exist in economies similar to that of the United Kingdom, and play much the same role there. Naturally, they are identified by different names, for example, Aktiengesellschaft (AG) in Germany, and société anonyme (SA) in France.

The word *Corporation* means the same thing as company and is often used in North America, where limited companies are said to be 'incorporated' and may put 'Inc.' after the company name. Another American usage is *Stock* instead of *Shares*. Even in Britain shares are traded on the Stock Exchange.

A company is defined by a *Memorandum of Association* which, among other matters, indicates its objects or aims. It operates under *Articles of Association*; these internal regulations define arrangements for general meetings of shareholders, powers of directors, means of altering share capital, winding-up provisions, and so on.

All companies registered in the United Kingdom are subject to the requirements of the Companies Act, which is revised by Parliament from time to time, usually to limit further the discretion of directors. Somewhat similar legislation exists to control companies registered in other countries.

Other Kinds of Company

Although this book will deal for the most part with large public limited companies that will be known to many readers, it is worth noting the existence of some other types. A *Company Limited by Guarantee* does not have a share capital; instead the members promise to contribute specified amounts if the company is wound up, that is, ceases to operate, leaving debts to be paid.

A *Close Company* is one over which five or fewer shareholders have effective control, by owning more than half the shares, or over which the directors have control. This situation is of significance for other shareholders, who lack even the theoretical degree of control that exists in a company with widely dispersed share ownership.

A *Private Company* is one that does not fulfil the conditions that define a public one. It cannot issue shares to the general public without first following carefully regulated procedures. The subsidiary and related companies controlled by large corporations are often close or private companies. An example is the Rover Group, now controlled by British Aerospace plc, with Honda Motors having a 20 per cent minority shareholding.

Finally, there is the *Unlimited Company*, whose members have an undefined liability for debts and other responsibilities. Such a company lacks the crucial feature which allows numerous individuals to pool their capital in a single enterprise without exposing themselves to incalculable financial consequences.

It is hardly an exaggeration to say that the Limited Company or Joint Stock Company is the foundation of the market economy with which we are familiar. Indeed, since such organisations appeared in the middle of the 16th century – among the first being specialised trading companies such as the Muscovy Company and the

East India Company – they have been the primary means of developing the world's economy.

Purposes and Responsibilities

The formal position set out above implies that the purpose of a company is simply generating profits for its owners, the shareholders, while complying with laws and regulations devised to protect its employees, customers and the community generally. Its wider responsibilities include dealing fairly and humanely with employees, honestly with customers and suppliers, and thoughtfully with the public and physical environment. Let us see how this prescription applies in two imaginary, but not unrealistic cases.

Consider first a company employing fifty people to manufacture envelopes in a modern factory in London. The company was founded by two of its present directors and has twelve shareholders, mostly members of the founders' families. The business is relatively simple and is well understood by the shareholders, who tour the plant annually and meet the employees then and on social occasions.

The prescription of purpose and responsibilities given above fits this imaginary company very well. Its customers are not dependent on this particular supplier; they need not buy its envelopes rather than those readily available elsewhere. Presumably working conditions in the modern factory are good; in any case, the directors are accessible to receive any complaints the employees wish to make. The failure of the company would not be catastrophic for its employees, since within London there are many other potential sources of employment.

Next let us think about a company with 350,000 employees, scattered through Europe, North America, Africa and Australia. The company was founded more than a century ago, and the descendants of the founders are now neither directors nor significant shareholders. The present directors are all professional managers, save for a few former politicians and a token woman. There are some 100,000 registered shareholders, mostly resident in Europe and North America, but 70 per cent of the shares are in the hands of a hundred corporations, mainly insurance companies, pension funds and investment companies. However, in no case does the investment in this particular company amount to more than 2 per cent of the assets of one of the large shareholders.

The company has numerous subsidiaries whose original shareholders were bought out many years ago. They manufacture a wide range of products, including pet foods, diesel and electric locomotives, medical equipment and highly toxic chemicals. The company also has subsidiaries that provide services, notably betting shops and insurance. In some towns – indeed in some small countries – it is the major employer and the only one utilising certain kinds of skill. In some regions it is the only convenient source of particular goods and services.

Although this second imaginary organisation operates within the same legal framework as does its much smaller brother, it has very different relationships with its owners, employees, customers, and the governments within whose jurisdiction it operates. Indeed, there is active debate on what these relationships actually are and even more on what they should be.

Example 1.1 Rolls-Royce plc 1989 and 1990: Shareholdings

This company controls a group that manufactures aero-engines and provides other engineering products and services. (Rolls-Royce cars are made by a quite separate organisation, a subsidiary of Vickers PLC.) After a number of years under the control of the British government, the company was privatised in 1987, that is, its shares were sold to the general public and to institutional investors. We shall consider the smallest and largest shareholdings as they were a few years after this flotation on the stock market.

On the 31st December 1989 the number of issued shares was around 959,848,500, of which the directors owned only 107,800. There were some 738,700 shareholders, but 527,300 of them owned 150 or fewer shares. More than one million shares were owned by each of 145 shareholders, mostly institutions such as pension funds, insurance companies and unit trusts. Together they owned 51.0 per cent of the company.

A year later, at the end of 1990, the shareholder register had shrunk to 640,100. Some 454,900 shareholders held 150 or fewer shares, while the 162 with more than one million shares held 56.6 per cent of the total shares in issue. This consolidation of holdings was to be expected following a privatisation process that created numerous small shareholdings. The Rolls-Royce share register is moving towards the pattern typical of large companies, with most of the shares held by institutions or corporate investors.

The directors state in their annual report that Rolls-Royce is not a Close Company. The structure of the share register supports this statement, which amounts to the claim that no small group of shareholders controls the company.

Note that the holder of 100 shares owns about one part in ten million of the company. How then does one assess his responsibility for its actions? Can he exercise any significant control over it?

1.2 The Business Environment

In the present context 'environment' means not only the physical environment, but all the organisations and people with which a company exchanges goods, services and money, and with which it interacts in less tangible ways. The *Providers of Capital* are of particular importance, as the source of funds to set up or to expand the business. They may be individuals, mutual societies, other companies, governmental agencies, or governments themselves. Some purchase the company's shares in the expectation of receiving periodic payments called *Dividends*. Others lend money in return for regular payments of *Interest* on the money lent, and the repayment of the loan in due course.

Fig. 1.1 represents the varied interactions between a company and the generalised environment within which it operates. The company considered is a manufacturing enterprise, the kind of organisation in which engineers are commonly employed. Its interactions with the world outside fall into the categories considered below.

Financial Transactions. These are the focus of interest in this book:

- Inward flows of money (solid arrows pointing in) from providers of capital, from customers in payment for the company's products, and sometimes from the government in the form of grants to encourage specific activities, and
- Outward flows of money (solid arrows pointing out) to employees, to suppliers of goods and services, to governments in the form of taxes and other levies, and to providers of capital, in the form of interest on loans and dividends on shares.

Goods and Services. It is to generate these transfers that the monetary exchanges are made:

- Inward flows of goods and services (broken arrows pointing in) from employees, suppliers and government, and
- Outward flows of goods and services (broken arrows pointing out) to customers, who may include governments.

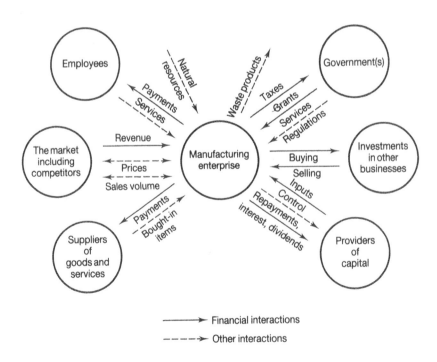

Figure 1.1   A commercial enterprise and its environment

Interactions with the Physical Environment. These attract intermittent waves of public concern, with a long-term increase in the general level of awareness:

- Absorption of natural resources such as air, water, land, minerals, forest products and petroleum, and
- Rejection to the natural environment of wastes such as heat, smoke, dust, slag, organic matter, water vapour, carbon dioxide, and other chemicals dissolved in water or carried in the air.

Mainly Intangible Interactions. These can sometimes be assigned a monetary value, for example, when shares are bought and sold, or when fines are levied:

- Transfer of a measure of control over the company to providers of capital, in return for the funds invested
- Inward and outward flows of money, shares and control, associated with the buying and selling of assets which may include complete or partial ownership of other businesses
- Regulations imposed by governments (or supranational organisations such as the European Commission) in whose jurisdiction the company operates, countered by such lobbying as the company can muster to influence legislators, and
- A complex interaction with the 'market', comprising both customers and competitors, which determines the quantity of goods sold, the selling price and, indeed, the nature of the products offered for sale.

Example 1.2 The General Electric Company plc 1990:
Major Financial Transactions

This is one of the largest UK-based manufacturing companies, providing a wide range of electronic, power and telecommunications systems, as well as consumer goods and medical and office equipment. We consider some of its major interactions with outsiders in the financial year that ended 31st March 1990.

For that year GEC reported:

- Sales of £8511 million, of which £3307 million were to customers in the United Kingdom
- Payments to employees of £1759 million, some indirect, relating to pensions and social security
- Taxes of £244 million, roughly equal sums going to the United Kingdom government and to overseas jurisdictions

- Dividends to shareholders of £220 million, and
- Net interest received of £129 million, this company being unusual in maintaining cash balances larger than its indebtedness, thus creating a positive balance of interest income and expense.

In the year in question GEC interacted with other companies in a dramatic manner. Its purchases of subsidiary companies cost £1028 million, while sales of subsidiaries brought in £315 million. The major events associated with these large monetary transactions were:

- Acquisition of The Plessey Group Limited by GEC and Siemens AG, the German electrical group

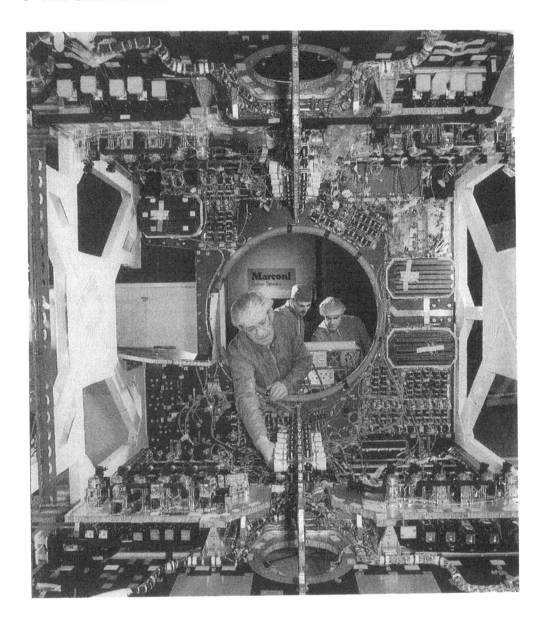

PLATE 1

Electronics payload for Eutelsat Communications Satellite from the GEC joint venture Matra Marconi Space NV.

- Acquisition of Ferranti Electronic Systems Limited
- Setting up GEC Alsthom NV, registered in the Netherlands, in partnership with the Companie Générale d'Electricité of France, and
- Setting up General Domestic Appliances Ltd, in partnership with the General Electric Company of the United States (no relation).

Of GEC's 107,400 employees, at the year end some 32,800 were reported as working overseas. As a consequence of the formation of the Joint Ventures (GEC Alsthom and GDA), the number reported as employed by GEC and its subsidiaries fell from 145,000 at the beginning of the year considered. It will be apparent that a large company like GEC is influenced by, and influences, the activities of numerous individuals, companies and even governments, around the world. The monetary values assigned to these transactions provide only a partial indication of the potential significance of the transfer of control over these large enterprises.

Since a considerable part of GEC's business is now carried out through Joint Ventures, partly owned by quite independent concerns, the relationship of the company with its markets does not have the simple form suggested in Fig. 1.1. In Fig. 1.2 the relationships with a single partner are indicated: the two companies collaborate in business sector A, compete in market C, and separately address other market sectors (B and D). Of course, each of the partners may be involved in other joint ventures, involving each other or different partners.

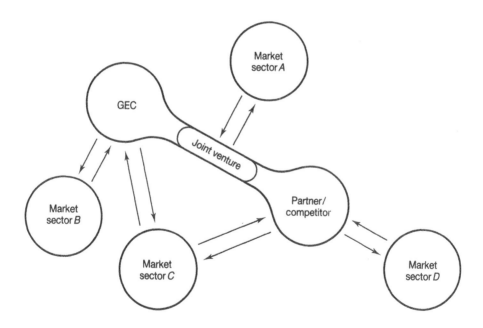

Figure 1.2 Interactions of The General Electric Company and a joint venture partner with various markets

1.3 Company Reports and Accounts

A condition for registration as a limited company under the Companies Act is that the Directors make an *Annual Report* to shareholders. This report must provide numerous specified pieces of information relating to the company's activities over the year. In fact, most companies go well beyond the legal requirements; they use the statutory report to provide shareholders and other interested persons with what is intended to be a polished description of the company and of the development of its businesses. Such reports are obviously intended to fulfil a public-relations role, in addition to letting the 'members' of the company know what the directors have achieved.

Companies also issue much briefer interim reports at the half-way stage in the financial year, and some report quarterly.

Organisations that are not strictly companies, for example, the Post Office, issue annual reports not unlike those of public companies. Presumably this encourages the staff to adopt a 'commercial' attitude and may possibly lead to a more efficient service.

Financial Statements

Annexed to the usually much-embellished Annual Report are the Accounts or, more precisely, Financial Statements for the period covered by the report. These seek to describe (or sometimes to mask) the financial transactions of the company during the period. The accounts are prepared in accordance with *Accounting Standards* laid down by bodies representing the accountancy profession. However, these 'standards' differ from country to country, and are under virtually continual review. Moreover, since there is great variety in the structure of companies and in their operation in different lines of business, directors are given considerable discretion in providing what they believe to be 'a true and fair view' of their activities.

The stream of annual reports and accounts issued by a company is the most coherent and consistent source of information on its activities. It is the feed-stock for the work of financial analysts, potential investors, the company's competitors, and even students seeking to understand company finance. A primary purpose of this book is to give insight into the nature, purposes and interpretation of the financial statements annexed to the annual report of the directors of a company to its shareholders.

The structure of the *Principal Financial Statements* will be considered in some detail in Chapters 3 to 5. At the time of writing they are:

- the Profit and Loss Statement
- the Balance Sheet, and
- a Statement of Source and Application of Funds.

    In 1992 the last of these is being replaced by

- a Cash Flow Statement,
    which is believed to provide a clearer view of the flow of money through the company. Some companies also provide

- a Value Added Statement,
    which shows how the 'wealth' or 'value' created by their activities is distributed.

As time passes, companies are being required by law and other regulation to provide more detailed and more highly standardised statements of their financial affairs. Moreover, the patterns of presentation adopted in different countries are becoming more alike. Nevertheless, there still exist significant differences in the form of the accounts prepared by companies registered in the United Kingdom, in the United States, and in other countries. For this reason we shall concentrate on the affairs of companies registered in the United Kingdom. In fact, some of these carry out much of their business overseas, and some have most of their employees overseas as well. The case of The General Electric Company, considered in Example 1.2, illustrates these points.

Environmental and Social Accounting

Truly comprehensive 'accounts' would incorporate every kind of interaction between a company and the world and its citizens. While conventional accounts deal only with the financial affairs of companies, many annual reports present a broader range of companies' activities.

Annual reports often include statements of ways in which companies serve their employees and the community more generally. Such statements are obviously intended to show that the issuing organisations are socially responsible and seek to meet society's needs in ways additional to their basic commercial activity. More recently, attention has turned to the environmental consequences of corporate activities. Some organisations provide *Environmental Audits* which indicate how they seek to improve their performance in this area. Chemical companies, such as Imperial Chemical Industries PLC, place great emphasis on the efforts they make to protect the environment.

It is interesting to speculate on whether the whole range of corporate activities could be represented in terms of monetary values. Social and environmental audits have not yet been merged into conventional financial accounting systems. Indeed, comprehensive environmental auditing would need to extend beyond the company itself, looking upstream to quantify effects in the supply chain, and downstream to examine all the benefits and drawbacks of the company's products.

Although they are not explicit, environmental charges do influence present-day accounts. In Fig. 1.1 solid arrows indicate monetary transactions, and broken arrows show transfers of goods, services and influence. Corresponding to nearly every solid arrow is an opposed dashed arrow; this suggests that an overview of the financial transactions also provides insights into the balancing interactions between a company and its generalised environment. There is an exception to this duality between finance and interaction: no direct financial consequences balance the interchanges with the natural environment, the use of resources and the discharge of wastes. To a growing degree governments, through taxation and regulation, provide surrogates for costs which nature, and individual members of the public, cannot impose themselves.

1.4 Graphical Representation of Company Finances

We now concentrate on the financial transactions described by conventional accounting methods. Fig. 1.3 shows schematically the monetary interactions identified in Fig. 1.1. The labels adopted for these transfers are those commonly used in company financial statements.

The boxes along the centre-line of Fig. 1.3 correspond for the most part to the interacting entities shown in Fig. 1.1. However, the enterprise and its employees have now been combined into a single Production Unit. This is in accord with the usual accounting convention.

Profit and Value Added

The items on the right-hand side of Fig. 1.3 appear in the *Profit and Loss Statement or Account*, the financial statement which shows how the annual profit of the company is derived. The *Sales or Revenue or Turnover* flowing in at the top is balanced by the varied expenditures and distributions below. When we consider the tabular form of the income statement, it will be helpful to look back at this graphical representation.

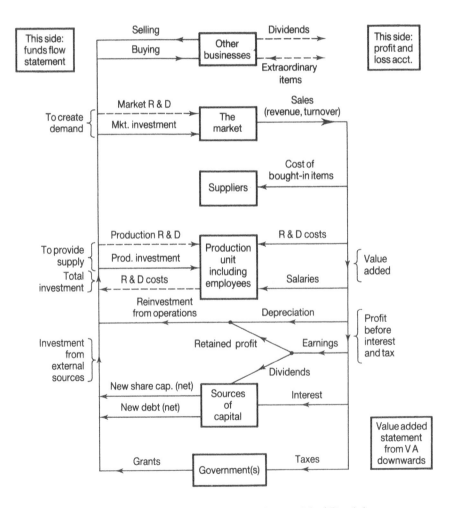

Figure 1.3 Identification of accounting qualities with the model of Fig. 1.1

In practice, the Profit and Loss Account of a company is more complicated than Fig. 1.3 suggests. Rather than pointing into empty space, the arrows at the top right-hand corner should feed into (or draw from) the profit stream further down the right-hand side of the figure. These complexities have been omitted to simplify the diagram.

The bottom part of the right-hand side displays quantities that appear in the *Value Added Statement*, another way of presenting a company's activities during the year considered. This starts with a basic trading surplus and shows how this cake is divided out:

- a slice to employees (usually much the largest)
- a slice to the taxman (not inconsiderable)
- a slice to lenders (interest)
- a slice to the shareholders (dividends), and
- a piece or two left on the plate for the company to digest internally (retained profit and depreciation).

The Flow of Funds and Cash

On the left-hand side of Fig. 1.3 appear the more important quantities of the *Statement of Source and Application of Funds* and of the *Cash Flow Statement* which is soon to replace it. This shows how the company acquired funds during the period, and how it utilised them, other than in running the production unit and its supporting services. This part of the diagram is not wholly driven by sales; additional capital may be introduced in the ways indicated in the bottom left-hand corner.

The other important financial statement, the *Balance Sheet*, is not represented explicitly in Fig. 1.3. This figure represents events during one accounting period, while the balance sheet lists the company's assets and liabilities at the end of the period. Thus the balance sheet gives the net result of all the transfers shown in Fig. 1.3, summed over all the years in which the company has been in business.

1.5 Investment and Reinvestment

The investment process is crucial to the creation and subsequent health of a business. From Fig. 1.3 we can learn a good deal about the mechanisms of investment and reinvestment in a company.

Starting Up. When a manufacturing business is first established, it has no sales revenue, and the right-hand side of the figure is blank. At this stage, investment depends upon the injection of funds by the Providers of Capital. These funds can, and indeed must, be used in two ways:

- to provide production facilities, that is the means of making goods for sale, and
- to establish a market, that is to design specific products and to a create a demand for them.

During this period the funds just invested also bear the varied expenses of the enterprise, such as salaries and costs of materials and services.

The Mature Company. Once sales revenue is generated, the other transfers of Fig. 1.3 come to life. After the running expenses have been deducted from revenue, something normally remains to be reinvested in the business. The total investment stream must be deployed in the same two ways:

- on production facilities, to increase the *Supply* of goods or to reduce their monetary cost and detrimental environmental effects, and
- on stimulating *Demand*, through the development of new products, market research, and sales support such as programmes to launch new products.

If a business is growing at a rate faster than that which its present turnover will support through the reinvestment of profits, further injections of capital will be required. For many years the computer industry was in this position, with rapidly rising demand pulling more and more fresh capital into that business sector. On the other hand, a company whose market is declining may become a cash generator, with a larger surplus than its present business can usefully absorb. The tobacco industry is a classic example of cash generation; in recent decades demand for its products has declined (in 'western' economies at least), while the business has remained quite profitable. Hence tobacco companies have had substantial surpluses to invest in diversification.

Research and Development. The monetary flows shown in Fig. 1.3 depart from accounting conventions in the representation of Research and Development (R & D). Accountants see expenditure on R & D as one of the costs of production, but in Fig. 1.3 it has been separated from other costs and shown as a form of reinvestment in the enterprise. This view of Research and Development activity is closer to the way engineers see it.

Expenditure on Research and Development is, like reinvestment in physical assets, devoted to two ends. It is used to improve the *Supply* of goods, for example, through the development of production methods that are more highly automated or that produce goods of higher quality. It is also used to stimulate *Demand* for the company's output, for example through the creation of new products or more attractive packaging.

Investments in Companies. At the top of Fig. 1.3 is a box labelled Other Businesses. A company may choose not to invest its available funds in existing activities, but in purchasing partial or complete ownership of other businesses. Companies in the tobacco industry, referred to earlier, have been able to do this because they commonly generate profits in excess of the need for reinvestment in their original business.

Companies that use their funds to purchase or set up subsidiaries may ultimately become *Holding Companies* whose business is managing subsidiaries and other assets, rather than manufacturing and trading themselves. The General Electric Company, considered in Example 1.2, is such a company. The combination of holding company, its subsidiaries and other related companies is referred to as a *Group*. It is the

Consolidated Accounts of the group, prepared by the holding company, that are usually made available to the public.

Of course, shares in businesses can be sold as well as bought, and the proceeds of sales are available to the company for any of the other purposes indicated on the left-hand side of Fig. 1.3. In difficult trading conditions peripheral assets may have to be sold to provide funds to sustain the core business.

Example 1.3 Glaxo Holdings plc 1990: R & D and Reinvestment

Glaxo, which is based in west London, is one of the great recent successes of British industry. It is now one of the world's largest manufacturers of prescription drugs, having over the years 1971 to 1990 achieved an average annual growth in sales of 16 per cent and in trading profit of 22 per cent.

For the financial year ending 30th June 1990 Glaxo reported Retained Profits of £464 million and Depreciation of £117 million. (Depreciation is a formal, not always realistic, estimate of the decay in value of a company's assets during an accounting period.) Thus the company's trading activities generated funds for reinvestment amounting to £581 million. This more than met its investment needs, and it did not raise significant amounts either of share capital or loan capital during the year.

Long-term success in the 'ethical' pharmaceuticals business depends upon the discovery of new drugs, the development of effective ways of making them, and a major marketing effort to get the medical profession to prescribe them. Worldwide, Glaxo employed 5700 people in Research and Development, one third of them in 'discovery research'; it spent £399 million on these activities. Hence the total amount that Glaxo reinvested in its core business was £980 million.

During this year Glaxo's capital expenditure on tangible assets amounted to £619 million, of which £233 million was spent on R & D facilities. New or expanded facilities were constructed in Italy, Spain, England and North Carolina. The total expenditure associated with R & D can then be reckoned as £632 million. This is 22 per cent of Sales and 45 per cent of the Trading Profit before deduction of R & D costs. These expenditures are far beyond those usual in engineering companies, where the margin between sales and costs is not as large.

Other capital expenditure was on new headquarters offices at Stockley Park, near Heathrow Airport, and on manufacturing plants in Scotland, England, Singapore, Taiwan, Indonesia and Spain.

Over the years Glaxo Holdings has divested itself of almost every subsidiary company not engaged in the core business of pharmaceuticals. It spends very little (relative to its size) on buying existing businesses, preferring to reinvest in its basic activities, which have proved able to absorb fruitfully much of the large surpluses that they generate.

2

What We Want to Do, and How

The preceding chapter introduced the subjects of this book: companies, their finances, interactions, growth and decline. We are now able to state the aims of this book and how it seeks to achieve them. As indicated earlier, engineers do not usually need the detailed knowledge of procedures that accountants acquire through years of study. We can set ourselves more limited targets.

2.1 Aims and Methods

The reader of this book should have two aims: acquiring knowledge and skills, and understanding how they apply to engineering businesses. It will be helpful if we can identify the major components of the body of knowledge and insights that we hope to establish.

Knowledge and Applications

The foundation for a discussion of company finance is an acquaintance with:

- the special language of finance, its vocabulary, grammar and syntax
- methods of reporting the financial operations of companies
- methods of assessing the financial condition of companies and other aspects of their operations, and
- the essential limitations of these methods of reporting and assessment.

The points listed above relate to the way in which a company reports its activities, rather than to the business itself. We must also know something about the processes that lie behind the records of financial transactions, such matters as:

- the general environment in which businesses operate, comprising customers, suppliers, other companies and governments
- corporate structures and relationships between the companies of a group
- ways of raising capital to finance business activities, and
- the interactions between companies, investors and governments, particularly as regards taxation.

Chapter 1 provided some of the necessary background, and the present chapter will explain more fully the accounting procedures used by companies to report their affairs and the thinking behind those procedures. If it were presented as a block, the background information would be difficult to digest. It has therefore been distributed through the rest of the book.

The body of factual material detailed above may not be intrinsically interesting, but its value becomes clear when it is applied to a number of commercially significant topics:

- factors influencing the growth, profitability and viability of companies
- links between risk, borrowing and the price paid for invested funds
- effects of inflation and exchange rates
- relationships between directors, shareholders, lenders and employees
- mechanisms and strategies for the control of companies, and
- commercial strategies applicable in the international marketplace.

The primary means by which we seek insight into these matters is by examining the recent experience of actual companies. It need hardly be pointed out that we cannot hope to explore these questions thoroughly in the space available.

Presentation and Learning

The limited goals that have been set are still quite challenging; the arrangement of this book is intended to help the reader to attain them. The opportunity will be taken to suggest some ways of coming to grips with the subject matter, that is ways of using the material in the book as a foundation for personal study and group work. This book does not provide examples for solution; instead it seeks to give readers the ability to create their own exercises, to develop skills and deepen understanding.

Modular Format. The text has been divided into coherent elements. When you want to refresh your memory, you should be able to locate topics easily using the index or the contents list at the beginning of each chapter.

Terminology. The study of business and accounting presents difficulties much like those of learning a foreign language. The special words of finance are defined as they arise in the text. You will notice that they are set in italic type. These terms are listed in the index, which will direct you to places where they are introduced. Sometimes a simple definition is provided when a concept first appears, to be amplified later, when further detail is required.

When struggling with the concepts of finance, you may find some comfort in views expressed by Philip Coggan in the *Financial Times* in September 1991:

> "Like all professions, finance is a conspiracy against the laity; it deliberately dresses itself up with jargon to make itself more complicated."

Case Studies. Many of the concepts and methods will be illustrated by reference to recent practice and experience in British industry. These examples amplify and reinforce the general points made in the text. Most relate to periods a year or so before this book was written, but that is usually not crucial to their applicability. Methods of financial reporting do change, but not so rapidly that these examples will not retain their relevance for some years. Moreover, industrial organisation and the problems of commerce remain much the same from decade to decade.

For the most part the examples concern companies active in one or more fields of engineering or technology. Students of engineering will already be familiar with some of these companies, and will need to know more about them as their careers develop. As was pointed out earlier, the success of engineering endeavours depends as much on the financial stability of the companies that carry them out as on the skills of their engineers.

Most of the case studies focus on methods of presenting and interpreting accounting information. However, they introduce broader aspects of corporate finance and development, and thus provide starting points for wider-ranging discussions of business strategy. Many of the matters touched on in these case studies will be reflected in the experience of other companies. A regular look through the financial sections of newspapers that cover such affairs will provide numerous examples of the kinds of situation analysed in this book.

Simplified Examples. While many concepts can be illustrated by reference to the activities and financial statements of real companies, it is sometimes expedient to construct simpler examples. This is especially true when considering aspects of borrowing, interest payments and taxation, which are inextricably woven together in actual sets of accounts. Moreover, the accounts of large groups, operating internationally, are superpositions of the accounts of numerous subsidiaries, many based overseas. Procedures that are essentially simple are applied in slightly different ways, to different periods of time, and in different currencies, and are then bundled together to create the accounts of the whole group of companies.

The assumptions adopted in constructing these idealised examples are clearly set out. It will be instructive to alter these assumptions, to see how the accounts respond and to investigate the sensitivity of the outcome to the initial postulates.

Annual Reports. There is no substitute for the real thing. The student using this book should acquire a few representative annual reports, if possible those of companies whose activities are of particular interest. Most corporate secretaries and public-relations departments are very willing to provide reports on request. An effective strategy is to apply the methods illustrated in the examples to other companies, using their reports as the source of information. You should also learn to find your way around a company report, and where to look for the most important pieces of information.

Example 2.1 Financial Terms

The special language of business and finance will be illustrated by passages taken from issues of *The Times* newspaper that appeared in July 1991.

Commenting on the interim (half-year) results of the Newman Tonks group, whose principal business is architectural hardware, but is also involved in property, *The Times* (*Tempus* 4 July 1991) reviewed this none-too-happy period as follows:

"Pre-tax profits in the six months to end April were just £7.3 million, 36 per cent down on last year . . . the benefits of Newman's acquisition programme were again hard to see . . . the balance sheet is looking rather healthier than the profit and loss account, thanks to last year's £32 million rights issue, which left the company ungeared at year end. Six months later . . . borrowings have risen to just under £20 million, giving a gearing of 25 per cent . . . the shares stand on a 3 p discount to the issue price of last year's new shares . . . the shares are on a price/earnings multiple of 13.5."

In discussing the 1990 results of Triplex Lloyd plc, a company that supplies the automotive and building industries, *The Times* (*Tempus* 17 July 1991) noted that:

"Operating profits from continuing businesses actually rose . . . The return on capital employed, excluding property profits, rose from 13.2 to 14.3 per cent ... and gearing has come down from 50 per cent to 38 per cent, clipping £1 million off interest charges. . . . A maintained dividend of 7 p is covered just 1.8 times by earnings of 12.7 p ... the County NatWest analyst expects £8 million profit . . . putting the shares . . . on a prospective p/e ratio of 9."

These newspaper reports were no doubt of considerable interest to investors in the two companies, and to those of their employees who could understand them. They contain more than a dozen technical expressions, mostly formed from common English words. The approximate significance of these phrases can often be guessed, but a precise understanding is sometimes necessary, and that requires systematic definition. What is the distinction, for example, between Operating Profits, Pre-tax Profits and Earnings? What is a Rights Issue? How is Return on Capital Employed calculated?

2.2 Principles behind Company Accounts

It is all too easy for engineers to be impatient with accountants. Engineers are tempted to contrast the supposed precision with which they describe the systems for which they are responsible, with the more arbitrary financial statements created by accountants. Such intolerance of the problems of another profession is perhaps natural. One of the purposes of this book is to increase engineers' understanding of accountants' responsibilities and difficulties.

Those preparing the accounts of a large group of companies have a formidable task. Taking British Aerospace plc as an example, we have accounts that seek to describe the financial affairs of a group that in 1990 employed some 128,000 people, its principal companies being registered in ten different countries on five continents (not to mention the Isle of Man). The activities of the group extended over aircraft (both civil and military), motor vehicles and civil engineering construction, and included joint ventures with French, German, Japanese and Swedish companies. It is not easy to describe an entire year's activities of this varied group in, say, twenty pages, and to report its financial affairs in a few columns of numbers.

Fig. 2.1 indicates the major influences that shape a company's annual report to its shareholders and the accounts that are appended to it. Some contribute mostly to the report itself, while others are responsible for the financial statements. We consider first the ideas behind the financial statements, turning to the report itself later in this chapter.

Standards, Policies and Concepts

Accounting Standards. The standards of particular relevance to most users of accounts are those relating to Disclosure, Presentation and Valuation.

On looking through a number of contemporary sets of accounts, one notices that they have much in common. This is because the accounting profession, with some support from legislation and international regulation, has laid down minimum standards for the *Disclosure* of financial information. Some companies reveal only what they must, others provide much more.

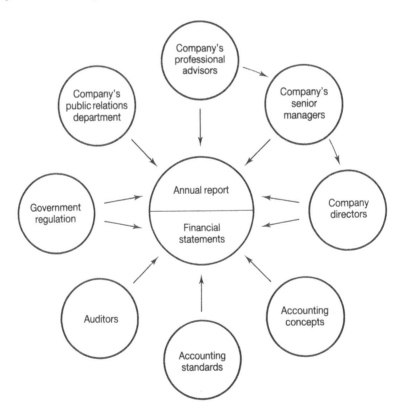

Figure 2.1 Contributors to a company's annual report and accounts

The *Presentation* of accounts is also becoming increasingly standardised, to the convenience of users. A company is now required, for example, to select the format for its financial statements from a small number of patterns.

There is still considerable variation in the methods adopted for the *Valuation* of assets. Some companies continue to value their land holdings at cost, that is, assign a *Book Value* unchanged since acquisition, possibly some decades ago. Others adjust the book value of property from time to time, usually on the advice of professional valuers, to reflect current market values.

Accounting Policies. Annexed to the financial statements that are the core of a company's accounts will be found a list of *Accounting Policies*, usually one or two pages in length. These indicate how the directors have dealt with the issues of Presentation and Valuation that were raised above, and cover such questions as:

- how valuations have been made
- how depreciation is calculated for different kinds of asset
- how the affairs of related companies are built into the accounts of the group
- how account is taken of companies that have joined or left the group during the accounting period
- how transactions in foreign currencies have been translated, and
- which of the recommended accounting standards have been adopted and which have not.

Such information may be of crucial importance in determining a company's financial position, but it is unlikely that many people other than professional accountants bother to read through these statements of policy, let alone grasp their implications.

Accounting Concepts. Although Accounting Standards and Policies develop as time passes, there lie behind them immutable Accounting Concepts. These fundamental ideas are reviewed below.

Accounts are prepared on the assumption that the company is a *Going Concern*, that is, will continue to operate for the foreseeable future. This implies, for example, that most of the goods in the process of manufacture will ultimately be sold for their intended purpose, rather than for scrap at a much reduced value, which would be more likely if the company were wound up. Another implication is that the company will continue to occupy the land on which its factories stand. Hence an up-to-date valuation of that land may be deemed unnecessary, as it is unlikely to be sold in the foreseeable future.

Another basic principle is that of *Consistency*. A particular company's accounts are expected to have the same form from year to year and to use the same rules of valuation. When changes are introduced, possibly as a result of new regulations, they must be clearly indicated. Moreover, it is customary to re-work the accounts relating to earlier years, and to present their key features again using the newly adopted conventions. A related concept is *Comparability*, the expectation that the accounts of different companies can be easily compared.

The concept of *Accruals* may seem unnatural to some readers. It deals with the problem of deciding when a sale has been made. Is it when the order is signed? when manufacture starts, is half completed, or is fully completed? when the invoice is prepared? when the product is dispatched? when the customer receives it? when the cheque in payment is received? when the cheque is cleared at the bank? The Accruals Concept deals with such problems by recognising revenues when they are earned and expenses when they are incurred, rather than when money changes hands. Thus goods are 'sold' when an invoice is issued, and goods are 'bought' when a firm order is placed. The application of this concept gives rise to an important distinction between 'profit' and 'net cash received'. A company can achieve a large profit (by making, dispatching and invoicing its products), but have a negative cash flow (by paying its employees and suppliers before receiving cash payment for the products it has 'sold').

Other accounting concepts are *Prudence, Objectivity, and Substance over Form*. Their intention is implicit in the names themselves. They are not always easy to put into practice, not least because the several concepts may conflict with one another. Finally, we note the principle of *Materiality*, which means that accounts need provide only information that is of significance to the user of the accounts.

Directors and Auditors

The Companies Act requires the directors of a company to present accounts that give 'a true and fair view' of the affairs of the business, but does not tell them how to do this. Increasingly, the discretion of directors is being limited by legislation, regulation and standards. Nevertheless, they still have considerable freedom in describing the financial condition of their company, and a corresponding legal responsibility, recognised by the requirement that two directors must sign the accounts to signify approval by the board of directors as a whole.

To balance the influence of the directors, independent experts called *Auditors* are employed to provide a report to shareholders, which is annexed to the Directors' Report. By law, company auditors must now be chartered accountants registered with a supervisory body, which has the right to assess their work. Behind the scenes auditors doubtless influence presentation and the accuracy of information, but sad experience shows that they cannot always protect shareholders against fraud, and they certainly offer no protection again bad business judgement.

For the most part auditors look at financial data, rather than at the objects and processes to which the numbers are supposed to relate. They usually confine themselves to making formal statements to the effect that they 'have followed Auditing Standards' and that 'in our opinion the accounts give a true and fair view of the state of affairs, properly prepared in accordance with the Companies Act'. Regrettably, and ironically, these statements relate to Form rather than to Substance. The major responsibility for the generation of meaningful accounts still rests with the directors and the professional staff they employ.

A Changing Scene

The recommended standards of Disclosure, Presentation and Valuation have changed over the years and remain under continual discussion. There are several agents of change:

- the accounting profession, which hopes to maintain a measure of responsibility for these standards
- the Stock Exchange, which seeks to maintain an orderly market in which buyers know what they are paying for
- governments, who seek to limit wrong-doing in this area, as in others
- supranational agencies, such as the European Commission, and
- the directors of companies, some of whom wish to provide a more comprehensive and comprehensible view of their companies' affairs.

As business activity, the work of accountants and the raising of capital all become more international in character, strong pressures towards standardisation of accounting conventions have built up. Adjustments now working their way through the world's accounting systems relate to such important matters as the way in which mergers between companies are represented, the treatment of financial subsidiaries, and the reporting of flows of funds and cash.

Some changes are not permanent. A decade or so ago inflation accounting was used extensively in the United Kingdom, since the double-digit inflation of that period cast doubt on conventional financial statements that took little account of it. *Current-Cost Accounting* was always contentious and was generally abandoned as inflation declined.

Of particular importance in the years to come will be the utterances of the recently established *Accounting Standards Board*. Compliance with its recommendations is compulsory, while the Accounting Standard Committee which it replaces had only an advisory role.

Example 2.2 Ferranti International plc 1989 and 1990: Misleading Accounts concealing Fraud

---

In 1988 Ferranti plc merged with the International Signal & Control Group PLC to form Ferranti International plc. Both companies produced electronic, computing and communications equipment, particularly for defence purposes. The marriage was not made in heaven, as the following extracts from the Chairman's Statement for 1989 indicate:

"The audited balance sheet of ISC at 31 March 1987 purported to show a net worth of $320 million, whereas we now believe that there was no net worth, and the profits for that and earlier years were substantially inflated."
"In summary, ISC's assets and profits were substantially inflated by a serious fraud which had been running for some years prior to the merger; the audited accounts of the ISC group

at 31 March 1987 did not show a true and fair view of that company's assets and liabilities; as a result, Ferranti was induced to enter into a merger it would not otherwise have contemplated and paid far too much for ISC."

Two extracts from the 1990 Chairman's Statement (from a new Chairman) carry the story forward:

"In November 1989, two of the Company's subsidiaries commenced legal action in England against the former chairman of ISC, Mr James Guerin, three other former ISC senior employees and five Panamanian companies which were used as vehicles for the fraud." (Judgements were rendered in favour of Ferranti in the amount of some $190 million, and steps were then taken to enforce them.)

"In January 1990, Ferranti International and its subsidiaries commenced legal proceedings for negligence against Peat Marwick McClintock, who were auditors to ISC at the time of the merger and became joint auditors to Ferranti."

This unfortunate case illustrates the limits on auditors' ability to verify claimed assets (in this case, contracts with overseas governments), and the ultimate responsibility of the directors of both the companies involved. The claim against the auditors was settled out of court with a payment of £40 million to Ferranti.

2.3 The Annual Report

The next chapter will consider in some detail the financial statements that usually accompany the annual report of directors to shareholders. Here we deal with the report itself. The upper part of Fig. 2.1 identifies those who assist the directors in the creation of their report.

As was pointed out earlier, many reports have a public-relations role, in addition to fulfilling statutory requirements. Indeed, more and more companies are offering to their shareholders an *Annual Review* instead of the full Report and Accounts, believing that most shareholders are not interested in the accounts, or able to make much sense of them. Such annual reviews contain only very brief financial statements, but seek to present a positive and attractive image of the company, using colourful charts and stylish illustrations. The companies recently privatised by the British government have special reason to introduce more compact annual reviews, as the privatisation process gave rise to very numerous small shareholdings.

From an international perspective the reports of companies registered in the United Kingdom are intermediate in the level of information provided on the company's financial and business operations. Some European companies offer much more. For example, the report of AB Volvo, the Swedish company with automotive and other interests, provides for each of its distinct business segments several pages densely packed with information. On the other hand, the report of the American Ford Motor Company does not even list its numerous large subsidiaries, although it does find space for numerous elegant photographs of its latest product range. The reports of overseas companies commonly give information on share-price variations during the year and describe key aspects of the financial performance quarter by quarter. Such information is not generally offered by companies registered in Britain.

The best way of becoming acquainted with company reports is to consider a representative example, and this is done below. The opportunity will be taken to say something about another statutory requirement imposed upon the directors of a public company, namely, the need to call an *Annual General Meeting*. From time to time *Extraordinary General Meetings* are necessary to deal with unusual business or business that cannot be deferred until the next AGM.

As one reads an annual report it may be helpful to have in the back of one's mind some thoughts expressed by Boyadjian and Warren in their book *Risks - Reading Corporate Signals*:

"Annual reports are written with a view of impressing the reader and it is worth a look to see what numbers the directors most wish to have their shareholders concentrate on."

Example 2.3 British Aerospace plc 1990: Annual Report and Annual General Meeting

This company is the largest in the United Kingdom in its field of activity which, in addition to military and civil aircraft, includes motor vehicles (the Rover Group), construction and property development.

The length of the 1990 Annual Report is 48 pages, of which 20 concentrate on financial statements and notes on them. This financial section is unrelieved by illustrations and unsupported by graphs, and cannot be described as welcoming.

Of the remaining 28 pages, five are devoted to the directors' statutory report to shareholders. This serves to:

- identify the present directors and indicate proposed changes and re-elections to the directing board
- list the shareholdings of directors and their families
- give some details of financial transactions between the company and individual directors

- propose that the auditors be reappointed
- give information on the company's share capital, including names of shareholders owning more than 3 per cent of any class of shares
- give details of proposed changes in the company's Articles of Association
- state briefly the nature of the company's business
- report the earnings for the financial year and the proposed dividends
- make a brief statement on the valuation of the company's real estate
- list charitable donations (and political donations, if any), and
- state employment policies, for example, in respect of disabled persons and other disadvantaged groups, and arrangements for share purchases by employees and directors.

Some Directors' Reports also give expenditure on Research and Development, but

PLATE 2

British Aerospace BAe 125 business jet.

British Aerospace does this elsewhere in its annual report. Somewhere the report must also provide information on significant 'post-balance-sheet events', that is, important transactions or calamities between the balance-sheet date and the publication of the report a few months later.

An annual report may include the agenda for the company's statutory Annual General Meeting. British Aerospace issues a separate agenda, with only a formal notice of the time and place of the meeting appearing in the report itself. The business of an AGM is fairly standard:

- to receive the accounts and the reports of directors and auditors
- to declare dividends
- to elect directors, and
- to appoint auditors and make arrangements to pay them.

It will be evident that the meeting is asked to approve the proposals set out in the Directors' Report.

Special business may also be transacted at the AGM. Typically this may:

- alter the Articles of Association
- authorise the directors to issue further shares or to repurchase the company's own shares, or
- introduce or alter share purchase or option schemes for employees or directors.

A Chairman's address is provided as an inducement to attend – another is sometimes to provide refreshments – but general meetings do not usually attract a significant fraction of those who have the right to attend. This is another indication that the control of shareholders over 'their company' is somewhat tenuous.

The 1991 meeting proved to be Professor Sir Roland Smith's last opportunity to address the shareholders of British Aerospace. In September 1991 he resigned as chairman, following a period of less than satisfactory trading. Apparently the non-executive directors played an important part

in this change of leadership. One of them, Sir Graham Day, took over as chairman on a temporary basis.

Turning to the glossy part of the annual report, we find a one-page summary of the year's results, illustrated with bar charts. This is followed by the chairman's two-page review of the year, which ends with an indication of the current trading position, that is, the position a few months into the year following the one to which the report applies. Such straws in the wind are taken very seriously by financial analysts, and it would be a serious misdemeanour for a chairman to misrepresent a company's position in such a statement.

A one-page tabular survey of the company's financial affairs over the preceding five years is usually to be found in annual reports. British Aerospace provides this, and supplements it with a two-page Financial Review, illustrated with graphs and bar charts. As 1990 marked the tenth anniversary of the formation of the company, the review looks back on its financial history during those ten years.

A page is devoted to Shareholder Information such as the number of individuals, the number of banks, the number owning more than 1,000,000 shares, and so forth. There is also a Financial Calendar for the coming year: when results will be announced, when dividends will be paid, and so on. Another page lists the Principal Subsidiaries and Associated Companies, indicating the percentage owned and country of incorporation. This is a surprisingly modest specification of the complex organisation whose activities the report purports to describe.

We are now left with 16 pages, and they are given over to a 'lively' Business Review. Here the print is large and well spaced. Even so, more than half the area is left blank or is devoted to rather uninformative photographs of the group's staff and products. Two pages are devoted to each of the company's four main business areas, two deal with R & D, two with international links and significant new developments, and two with the group's employees and contributions to the community.

The topics treated in this Business Review

are those found in most present-day reports, but the detail falls short of that commonly made available. Some readers might have hoped for information on:

- the finances of Airbus Industrie, a French-registered Groupement d'Intérêt Economique (a kind of private company), in which BAe has a 20 per cent interest
- the activities of Balast Nedam, a civil engineering company whose results have been combined with those from BAe's property enterprises, and
- the Rover Group, for example, the number of vehicles produced and the position of Land Rover Limited, a distinctive subsidiary.

Many of the group's businesses are not mentioned at all in the report, save for their appearance in the list of related companies. Nor is much provided by way of analysis of the group's financial performance, or of the position of its operating companies in their particular markets.

One wonders why British Aerospace's reporting of its activities is relatively thin. Is this apparent secrecy the consequence of the company's long connection with the defence industries? Has it elected to shield its 'owners' and other readers from the true complexity of its business and financial affairs? Does this vagueness spring from some combination of these motives? Whatever the reason, the reader will search in vain for key pieces of information about the British Aerospace group.

3

The Financial Statements

3.1 Consolidated Accounts of a Group

On looking through the accounts of a large company, one encounters a page headed 'Consolidated Balance Sheet' or possibly separate balance sheets marked 'Group' and 'Company'. Indeed, most of the financial statements are labelled with the adjective *Consolidated*. We need to understand what this means.

The greater part of the report and accounts issued by a public limited company relates not to that single organisation, but to the affairs of a *Group* of companies that it owns wholly or partially. The 'group' or 'consolidated' statements subsume the financial affairs of all of these companies. The Company Balance Sheet for the reporting company, which is referred to as the *Holding Company* or *Parent Company*, includes as assets the value of its shares in the other members of the group. Some of these companies may themselves be public limited companies, with their shares traded on stock markets, and issuing their own reports to shareholders.

The companies of a group are classified according to the level of control that the parent company has over them, as indicated below. The finances of these classes of holdings are built into the group accounts in different ways, which will be explained when we look at the statements in detail in Chapters 4 and 5.

A company is called a *Subsidiary* if the holding company owns more than 50 per cent of the voting shares, or in some other way controls the membership of the board of directors.

An *Associated Company*, which may also be described as a *Related Company*, is one over which the parent has a significant influence. This is commonly interpreted as having control over 20 to 50 per cent of the voting shares. If there are no other share-holders with holdings of this order, the degree of influence can be very considerable.

The term *Joint Venture* is often applied to an associated company established with a partner or partners for a specific purpose. For example, the consortium responsible for the construction of the Channel Tunnel, Transmanche-Link, was created for that specific purpose, and will be wound up when the project is finished. Other collaborations continue for many years, for example, Airbus Industrie.

Holdings below 20 per cent of the issued shares are termed *Investments* and, if the parent does not seek to influence the companies, may be described as *Trade Investments*.

Fig. 3.1(a) shows the assets of an imagined group of rather simple structure. If the parent does not trade itself, but acts only as a holding company, its operating assets may be quite modest. In addition to their investments in businesses, the group companies will maintain a float of more liquid assets to support their trading and investment activities.

Fig. 3.1(b) illustrates a more complex structure, in which intermediate holding companies control some assets, while others are owned directly by the parent com-pany. Some of the group's operations fall within *Divisions* that organise its activity in specific industrial sectors or geographical areas.

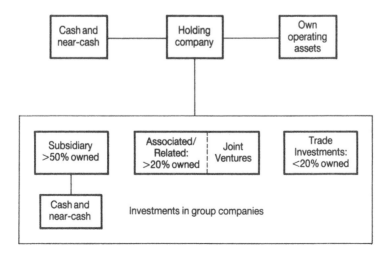

Figure 3.1a   Asset structure of a simple group of companies

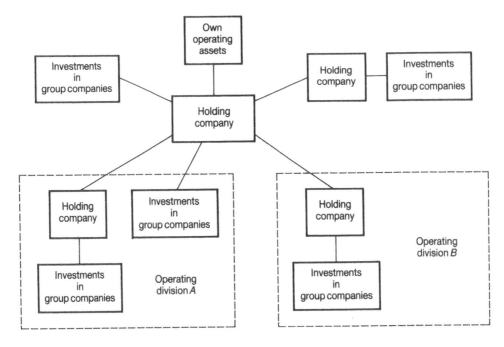

Figure 3.1b   A more complex asset structure

Example 3.1: GKN plc 1990: Structure of the Group Controlled

GKN - the name is a contraction in the modern fashion of Guest, Keen and Nettlefold - had its roots in the steel industry. In recent decades GKN has moved away from that industrial sector, to establish a strong position in automotive components, where it is one of the world's more significant players. It has also diversified into other aspects of engineering and into industrial services. The following statement, taken from the holding company's report, tells us something about the group controlled by GKN and the way in which it exercises control:

"The issued share capitals of the 131 companies which at 31st December 1990 comprised the GKN group are held directly (*) or indirectly by GKN plc through intermediate holding companies which are registered in England, Netherlands, USA, Germany, Australia and South Africa. Certain intermediate holding companies do not prepare consolidated accounts."

In fact, only four of the group companies listed in the report are marked (*); the great majority are controlled through the 'intermediate holding companies'.

The businesses of the GKN Group are organised into three Divisions, each with its own managing director:

● Automotive Drive Line Systems
● Engineered and Agritechnical Products
● Industrial Services

This operating structure serves to coordinate the activities of the numerous legally distinct companies of the group. A significant part of the Industrial Services Division

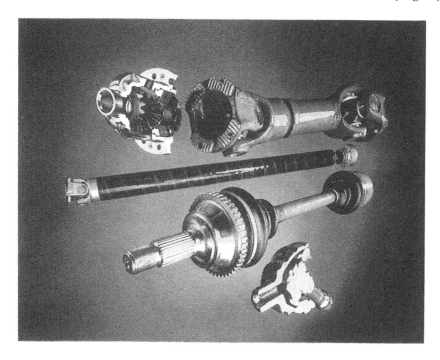

PLATE 3

Automotive drive-line components manufactured by GKN.

is controlled jointly with Brambles Industries of Australia; the companies involved are mostly 'associates' of GKN plc.

GKN plc has at present very little investment in shareholdings so small that they fail to give it a significant influence. However, three associated companies fall outside the divisional structure. One is the helicopter manufacturer Westland plc (29 per cent owned by the GKN group); there is a general expectation that GKN will seek to acquire control at some time by purchasing further shares. A second associate is United Engineering Steels Ltd (39 per cent owned by GKN), which is controlled jointly with British Steel plc. Here the future involvement of GKN is likely to be quite different: it may well complete its withdrawal from the steel industry by disposing of this investment.

The turnover of GKN's subsidiaries amounted to £2040 million in 1990, while its share of the sales of associated companies was £556 million. Thus it appears that the total turnover of the associated companies exceeded £1000 million.

3.2 The Principal Statements

Before turning to the financial statements themselves, we consider the period of time to which they relate. Normally, the accounts that accompany an annual report relate to a 52-week period, not necessarily coincident with a calendar year. However, a company sometimes changes its reporting date, possibly to bring it into line with other companies of the group. The more general term *Accounting Period* is the time between any two asset valuations, but for convenience we shall generally refer simply to the 'year' to which the accounts relate.

To help readers assess the company's development over the year, accounts reproduce much of the data relating to the last reporting period. Those data may be adjusted, to allow for information not available at the earlier date, or for modifications of accounting conventions.

What are the Principal Statements?

Present-day accounts issued by companies registered in the United Kingdom are based on three statements: the Profit and Loss Statement, the Balance Sheet, and the Statement of Source and Application of Funds. (As was noted earlier, the last is soon to be replaced by a Cash Flow Statement.) A string of notes is appended, presenting amplifications and explanations in a fairly standard way. The reasons for this degree of standardisation were explained in Section 2.2.

The *Profit and Loss Statement or Account* may be given one of the alternative names *Statement of Income or Earnings*, particularly in accounts issued by overseas companies. As is evident in Fig. 1.3, this statement shows how the year's Turnover (or Sales or Revenue) is translated into the annual Earnings of the group, following disbursements during the year to suppliers, employees, taxation, providers of capital, and so on. We shall find that, although this statement seeks to define the income achieved over the past year, it is conditioned by judgements about the future value of assets and future costs and obligations.

The *Balance Sheet*, which may be called the *Statement of Financial Position*, lists the assets and liabilities of the group on a particular date, namely, at the end of the accounting period. It demonstrates also that the sum of asset values equals the sum of the liabilities. Among the 'liabilities' is shown *Shareholders' Funds* (often abbreviated to SHF); an alternative name is *Equity*. As a consequence of the balance imposed within this statement, the SHF equals the *Net Value* that appears on the assets side of the Balance Sheet. It may seem odd that SHF is classified among the 'liabilities', until one recalls that the company is a legal entity distinct from its owners, the shareholders. If the company is wound up, the directors seek not only to pay off its debts, but also to return to the shareholders their proportions of the remaining value.

The *Statement of Source and Application of Funds* also has alternative names: *Statement of Funds Flow* and *Statement of Changes in Financial Position*. Its nature is evident in Fig. 1.3, which shows that this statement records all the funds coming in during the accounting period and indicates where they go. Consideration is given both to funds generated from operations and to funds invested in (or withdrawn from) the business.

Algebraic Relationships

For numerate people like engineers, the way in which these financial statements fit together is clarified by expressing them algebraically.

We start with the Balance Sheet. The end-of-year Total Assets A must equal the sum of the Liabilities Proper L plus the Shareholders' Funds S. The SHF can be broken down into:

- *Share Investment I*, the funds paid in over the years by the purchase of shares, and
- *Retained Profits R*, the sum of all the profits that have been kept within the company over its years of operation.

Using these symbols we can set down the identities:

$$A = L + S = L + I + R \tag{3.1}$$

Turning to the Profit and Loss Statement, we note that the Retained Profit for the period is derived from the Turnover T by deducting Internal Expenses E, Taxation G and Dividends D. Hence

$$\Delta R = T - E - G - D \tag{3.2}$$

This increment is the primary means by which the results of the year's activities are fed through to the Balance Sheet.

A basic form of the Funds Flow Statement can be derived from the balance-sheet identity of Equation (3.1):

$$\Delta R + (\Delta I + \Delta L) = \Delta A$$

This states that the net change in assets during the year balances the funds flowing from the business (ΔR) plus the new capital injected ($\Delta I + \Delta L$) by shareholders and lenders.

A more complete representation of the funds made available during the year can be obtained by identifying E_d, the *Depreciation* component of the total expenses E. This is a notional expense introduced to spread the cost of acquiring major assets over a number of years. Now we can write

$$\begin{aligned}(\Delta R + E_d + G + D) + (\Delta I + \Delta L) \\ = (\Delta A + E_d) + G + D\end{aligned} \tag{3.3}$$

On the left-hand side of this equation are collected the sources:

- all the funds flowing from trading activity (the first bracket), and
- all the new capital inputs (the second bracket).

The total funds made available are balanced by three ways of 'applying' funds:

- in buying assets ($\Delta A + E_d$)
- in paying taxes G, and
- in making a dividend pay-out D.

The term *'Cash Flow'* is sometimes used for funds generated from operations, the quantity $(\Delta R + E_d + G + D)$; we shall see later (Section 3.4) that this does not accurately measure the net cash generated by the business over the year.

The three equations developed above indicate in a general way how the basic statements are related. However, when we examine real accounts, we will encounter complexities that cannot conveniently be introduced here. Allowance needs to be made for *Minority Interests*, that is, the proportions of earnings and assets attributable to minority shareholders in the group's part-owned subsidiary companies. Also, changes in exchange rates alter the values of assets as measured in the reporting currency, and the financial statements must reflect these effects. Finally, the adoption of the Accruals basis of accounting means that anticipated income and expenditure for the accounting period will not equal the actual cash transfers during the period. We shall see, for instance, that the taxation element identified by the symbol G in Equation (3.3) is generally not the same as that appearing in Equation (3.2). In the latter, G is tax on the year's profits that will have to be paid sometime. In equation (3.3), G is the actual tax payment made during the year to which the report refers.

Example 3.2 Ford Motor Company Limited 1989: Abbreviated Financial Statements

---

The Ford Motor Company registered in England is now a wholly owned subsidiary of the Ford Motor Company of the United States. In the past it had other shareholders, but they have been bought out by the parent company.

For some years Ford has made more cars than any other British automobile company, and its products will be familiar to every reader. It is itself a holding company, with such subsidiaries as Jaguar Ltd, Aston Martin Lagonda Ltd and the Ford Motor Credit Company Ltd, and associated companies such as Iveco Ford Truck Ltd.

Ford UK issues an annual report and audited accounts much like those of a company whose shares are quoted and traded. The basic features of these accounts will be used to illustrate the typical structure of the financial statements. All the monetary values given in Tables 3.1 to 3.3 are in £ million. The symbols of Equations (3.1) to (3.3) are appended to relate the figures in the tables to the concepts introduced above.

Table 3.1   Group Profit and Loss Account

Turnover (T)	6732	
Expenses (E)	(6249)[a]	
Profit before Tax		483
Taxation (G)	(76)	
Profit after Tax		407
Minority interests[b]	4	
Earnings ($D + \triangle R$)		411
Dividends (D)	(451)	
Retained Profit ($\triangle R$)		(40)[c]

[a] Brackets are conventionally used in financial statements to identify quantities to be subtracted.

[b] This line represents the proportion of the year's Earnings attributable to minority shareholders in subsidiaries, where Ford holds the majority of the shares. Usually this element is negative, thus allocating a fraction of the subsidiaries' profits to the minority holders. The positive contribution shown here could result from an over-estimate of losses by those subsidiaries in earlier years.

[c] In this year the Dividends exceed the Earnings, so that the Retained Profit is actually negative.

Table 3.2   Group Balance Sheet

Fixed Assets[a]	3298		
Current Assets[b]	4514		
Total Assets (A)			7812
Current Liabilities	5578		
Long-term Liabilities	1262		
Total Liabilities (L)[c]		6840	
Share Investment (I)	39		
Other Reserves $\left.\begin{array}{l}\\\\\end{array}\right\}$ (R)[d]		14	
Profit and Loss Account	886		
Shareholders' Funds (S)		939	
Minority Interests[e]		33	
Total Liabilities including SHF			7812

(a)   Fixed Assets usually comprise land, machinery and other tangible items. Ford UK has included a significant 'intangible' element (£1221 million), representing much of the value of subsidiaries purchased during the year.
(b)   Current assets include cash and securities that can be converted into cash quite quickly.
(c)   Current liabilities are those likely to be paid off within a year. Long-term liabilities are those, like much debt, with a longer term.
(d)   The Profit and Loss Account is the main repository of accumulated annual retentions. Other Reserves contain retained profits that are not generated by ordinary trading. Reserves do not represent cash, which is within the Current Assets; indeed, they appear on the Liabilities side of the Balance Sheet.
(e)   This is the fraction of the SHF of the Group that has been contributed by those with Minority Interests in group subsidiaries.

Table 3.3   Group Source and Application of Funds

Profit before Tax ($\triangle R + D + G$)	487		
Depreciation E_d[a]	251		
Adjustments[b]	(12)		
Funds from Operations (or 'Cash Flow')		726	
Increase in Liabilities ($\triangle L$)	2496[c]		
Increase in Share Investment ($\triangle I$)	0[d]		
Total Capital Input		2496	
Total Funds Made Available			3222
Investment in Subsidiaries ($\triangle A + E_d$)	1520[e]		
Other Assets Purchased	1010		
Tax Paid (G)	241[f]		
Dividends Paid (D)	451		
Total Applications of Funds			3222

(a)   As Depreciation is a notional expenditure, representing past and possible future costs, it is part of the trading surplus of the year considered and is available for application.
(b)   Not worth worrying about at this stage.
(c)   Much debt was incurred to fund the purchase of Jaguar plc (now Jaguar Ltd).
(d)   No shares in Ford UK were issued or cancelled.
(e)   Jaguar was the largest purchase.
(f)   This is larger than the Tax item appearing in the Profit and Loss Account above. The year 1988 was more profitable and attracted more tax, which was actually paid in 1989.

Taken together, these statements tell us some important things about Ford's UK operations:

- Although the company was very profitable in 1989, its large dividend meant that the profits were not retained in this part of the Ford organisation.
- Only a small fraction of the total assets are 'owned' by the single shareholder (although it controls them); most of the capital has been provided by lenders.
- The 'Funds Flow' or 'Cash Flow' of the company was dominated in this year by asset purchases and a corresponding increase of debt.
- As a consequence of the small deduction of negative Retained Profits, the SHF fell to £939 million from £979 at the end of 1988. If the much larger effects of the purchase of subsidiaries had been represented in the accounts in a more conventional manner, the SHF would actually have been negative at the end of 1989.

3.3 The Value Added Statement

Although the three statements considered above are currently the 'principal' elements of most accounts, other ways of looking at the year's activities provide additional insights into a company's financial operations. One of these is implicit in Fig. 1.3, namely, the *Value Added Statement*. This shows the proportions of 'the wealth created by the company' that have been transferred to various beneficiaries, who are sometimes called *Stakeholders*. This term is a generalisation of Shareholders, and indicates a supposed responsibility to a community wider than the nominal owners. The philosophical justification of this position is a matter for debate.

To express the Value Added Statement algebraically, we break down the company's (or more properly the group's) expenses for the year:

$$E = E_d + E_e + E_i + E_m \tag{3.4}$$

where E_d is the Depreciation, introduced earlier
E_e represents the costs of Employment
E_i represents net payments of Interest to lenders
E_m represents the costs of Materials and Services.

The profit and loss statement, Equation (3.3), can now be expressed in the form:

$$\text{Value Added} = T - E_m$$
$$= E_e + E_i + G + E_d + \Delta R + D \tag{3.5}$$

Since this is really another way of writing the profit and loss statement, it does not provide essentially new information. Nevertheless, it gives useful insights into the interaction of the company with the various 'stakeholders' or beneficiaries from its activities.

Measures of Retentions

Students often have difficulty in distinguishing between the entities called Value Added, Cash Flow, Profits, Earnings and Retentions. All relate in some way to the ability of a company to make money. Why are there so many of them? How are they related?

The diagrams of Fig. 3.2 show how each of these measures of *Retentions* is built up from elements of the profit and loss statement. Along the bottom of this figure are shown the 'beneficiaries'. To indicate the relative magnitudes of the benefits they receive, values for the 1989 operations of the BOC Group plc are shown. BOC is one of the world's largest manufacturers of industrial gases, and also makes health-care and high-vacuum equipment.

From the point of view of the whole community, it is the Value Added that measures the benefits of the company's activity. The 'Cash Flow' measures its ability to sustain its ongoing business; this quantity is discussed more fully in Section 3.4. Earnings (or Profit after Interest and Tax) measures the gains of the owners of the company. Finally, the company itself makes Retentions to provide for its long-term development.

Expenditure on Research and Development is not usually extracted from the body of Cost of Sales, but is buried among the employment costs and payments to suppliers. However, it is realistic to include R&D costs as part of the total re-investment, as indicated on the right-hand side of Fig. 3.2.

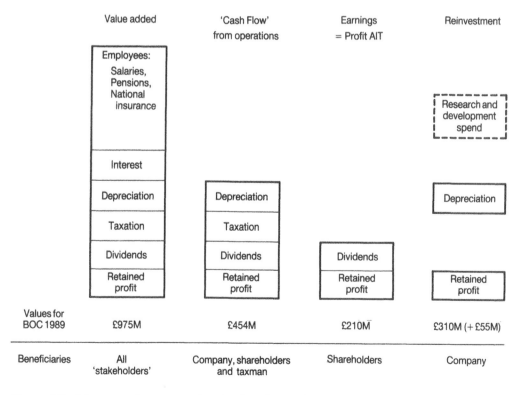

| Beneficiaries | All 'stakeholders' | Company, shareholders and taxman | Shareholders | Company |

Figure 3.2 Measures of a company's retentions from sales

None of the quantities represented in Fig. 3.2 need be equal to the change in the group's holdings of cash or equivalent liquid assets. The reasons for this will be explored in Section 3.4.

Example 3.3 The Ladbroke Group plc 1990: A Comprehensive Value-Added Statement

---

The main businesses of this international group are hotels (Hilton), off-track betting, property, and retailing (Texas Homecare). It publishes a particularly comprehensive value-added statement, and this is the reason for considering it here – although even engineers stay in hotels and engage in DIY. The statement issued by the company is given in slightly altered form in Tables 3.4 and 3.5; all values are in £ million.

Table 3.4   Generation of Value Added

Turnover		3800.5
Cost of goods sold, food, drink, services, betting-winnings, etc., all net of VAT	(2511.4)	
Extraordinary Items	(13.5)	
Group Value Added		1275.6

Table 3.5   Applications of Value Added

To employees:			
Take-home pay, company pension contributions, etc.		303.2	
Employee Share Scheme		1.5	304.7
To central and local government and other agencies:			
Betting duties and levy	—UK	203.2	
	—Overseas	33.2	
Rates, VAT, PAYE and NI deductions, etc.	—UK	113.4	
	—Overseas	112.1	
Corporate Taxation	—UK	48.7	
	—Overseas	31.1	
	—Adjustmts	(3.6)	538.1
To providers of capital:			
Interest to lenders		154.5	
Dividends to shareholders		91.5	246.0
To minority shareholders in subsidies:			0.4
Retained in the business:			
Depreciation		62.4	
Undistributed profit		124.0	186.4
Total Applications			1275.6

The reporting of transfers to governments, both in the UK and overseas, is especially instructive. The levies by local governments (rates), and payments of value-added tax (VAT) and betting levies, are usually submerged within the broad scope of cost of sales. Ladbroke has teased out these components of costs and has distinguished transfers to the British and other governments. Hence we can form a clearer picture of the way in which the 'wealth' is disbursed.

In this presentation governments appear as much the largest beneficiaries of the group's activities; they receive 42 per cent of the value added. The corporation tax on the group's profits is a modest component of the total governmental levies. Employee salaries and other benefits (some to be taxed when ultimately received) come next, with 19 per cent of the cake. The shareholders receive as dividends only 7 per cent of the value added, though their share rises to 17 per cent when retained profit is included.

3.4 Cash Flow

These words are used in two rather different ways, and we need to understand both. One use of the term emerges from the algebra of Section 3.2. The quantity within the left-hand brackets of Equation (3.3) is loosely called 'Cash Flow'; we have seen that this represents the 'funds' made available by the year's trading. The label 'cash' is not really justified, since the adoption of the Accruals approach means that the profit calculated for the year may be very different from the cash or ready money generated by the company. Acquisitions and disposals of subsidiaries further confuse the relationship between the generation of 'funds' and the cash available at the bank.

For an established group operating more or less in equilibrium, without major changes in group structure or other upsets during the year, this loosely defined 'Cash Flow' provides a useful measure of the cash actually generated within the business. However, the circumstances described are those in which we are least likely to be concerned about a company's ability to generate cash to pay its bills.

It is the second and better-defined use of the term Cash Flows that is of more interest here. These are monetary transfers that contribute directly and unequivocally to the actual cash available at short notice. The net result of all such transfers gives the increment (from year-end to year-end) in the balance-sheet entry labelled Cash. (Other possible labels are Deposits and Cash, and Cash at Hand and in Bank.)

American accounting practice has been ahead of British practice in this area. For some years American companies have been required to provide a *Statement of Cash Flows* or *Statement of Changes in Cash Position*. The parent Ford company in the United States provided such a statement for 1989, unlike Ford UK, whose compressed Funds Flow Statement is shown in Example 3.2.

A Cash Flow Statement determines separately the cash that is generated (or absorbed) by operations, by financing activities, and by investing activity – and, of course, the algebraic sum of the three contributions. Many of the entries in this statement do not occur in the profit-and-loss and funds-flow statements. Hence we cannot readily represent the Cash Flow Statement using the symbols of Equations (3.1) to (3.5). This technical difficulty points to a major virtue of the cash flow statement: it provides information that cannot be gleaned from the others.

We have been considering a retrospective cash flow statement, giving the net results of the entire past year's trading. A company must also keep track of its cash balance throughout the year, to monitor its ability to pay bills as they come due. Such a *Cash Flow Forecast*, predicting the cash that will be available at every point during the coming year (or an even longer period), is an important tool of management. Only by constructing such a forecast can one be confident that the complex series of receipts and outgoings will not run the company short of cash at some time.

Example 3.4 British Steel plc 1990/91: Statement of Cash Flows

---

This company, which should need no introduction, has numerous American shareholders, and therefore provides 'Supplementary Information for North American Investors'. This includes the statement presented in compressed form in Table 3.6.

Here 'cash equivalents' means cash plus investments that can be converted to cash within three months. Note that bracketed quantities are negative, and that a negative increase is a reduction.

This statement gives a clearer picture of the balance between the three ways of acquiring cash (trading successfully, selling assets, and raising fresh share or loan capital) and the three ways of using cash (running the business, buying assets and repaying providers of capital). It would be instructive to know also how the purchase of assets is split between simple replacement of plant and equipment, and expenditure on expansion and genuine improvements. Such information is not usually made available.

For comparison, Table 3.7 gives measures of profit and 'cash flow' extracted from other parts of the accounts for the same accounting period. None is as large as the Net Cash from Operating Activities given in Table 3.6, although the 'Cash Flow' before

Table 3.6   Consolidated Statement of Cash Flows (for year to 30 March 91, values in £ million)

Cash flows from operating activities		
Cash received from customers	5371	
Cash to employees, suppliers	(4496)	
Dividends and interest	144	
Interest paid	(23)	
Income taxes paid	(121)	
Unusual items	(91)	
Net cash from operating activities		784
Cash flows from investing activities		
Proceeds from sale of assets and loan repayments	33	
Purchase of assets	(455)	
Purchase of subsidiaries (net)	(450)	
Net cash from investing activities		(872)
Cash flows from financing activities		
Net increase in borrowing	(189)	
New borrowings	132	
Sale of shares	0	
Dividends paid	(170)	
Net cash from financing activities		(227)
Exchange translation effects		1
Net increase in cash equivalents		(314)
Cash equivalents at start of year		982
Cash equivalents at end of year		668

the exceptional item (BEI) is not very different. Note too that none of these four measures suggests that the cash readily available to British Steel fell by more than £300 million over the year. There seems to be little reason to worry, however; the cash balance remained above £650 million.

The conventional Funds Flow Statement provided by British Steel indicates a reduction in 'net liquid funds' of £139 million. This is less than half the decrease calculated in the Cash Flow Statement of Table 3.6, which is directly reconciled with the cash to hand at year end.

We conclude that the Statement of Cash Flows does give insights beyond those provided by the principal financial statements that have been conventional in the United Kingdom.

Table 3.7   Other Measures of British Steel's Results

Profit before exceptional items (mostly redundancy costs)	471
Profit before taxation	254
'Cash Flow' (Profit BEI + Depreciation)	707
'Cash Flow' (Profit BT + Depreciation)	490

3.5 The Notes on the Accounts

In a typical set of accounts, the basic financial statements occupy four pages or so, and are followed by some twenty to twenty-five pages of Notes on the Financial Statements. Those provided by UK-registered companies follow a fairly standard pattern, and a good deal can be learned about the whole family by examining a representative set of notes. It is convenient to look at those of British Aerospace, since that company's annual report and accounts have already been described in Example 2.3.

Example 3.5 British Aerospace 1990: Content of Notes to the Accounts

Some of the notes were considered in Example 2.3, as they form part of the directors' statutory report to shareholders. Among the notes we find details of the company's share capital, including the shareholdings of the directors, and information on the emoluments of the directors (that is, what they have decided the company should pay them). Mention has been made of the Accounting Policies used in preparing accounts (Section 2.2); those adopted by British Aerospace are given in Note 1.

Many of the other notes are expansions of individual lines of the balance sheet and the profit and loss statement, but the detail is unlikely to add much to the understanding of most of those who receive the report. Notes likely to be of greater interest are those relating directly to the company's business or to the intrinsic value of the company. We consider some of these below.

Note 2: *Sales and Trading Profit*. British Aerospace breaks down its business into the six sectors or segments shown in Table 3.8. A geographical analysis is also provided to meet stock-exchange requirements; this is shown in Table 3.9.

Table 3.8 Analysis by Business Segment
(values in £ million)

	Sales	Trading Profit
Defence systems	4423	486
Commercial aircraft	1560	35
Motor Vehicles	3785	55
Property, Construction	577	19
Space, Communications	120	(12)
Other	75	2
	10540	585

Table 3.9 Geographic Analysis of Sales
(values in £ million)

	1990	1989
United Kingdom	3668	3496
Middle East	2801	2057
Europe	2440	1905
US/Canada	757	971
Far East	501	308
Central, South America	151	49
Australasia	129	195
Africa	93	104
	10540	9085

Anyone with the slightest curiosity about this company will want to know which of its business segments are the more important and the more profitable, and whether it is heavily reliant upon sales in one part of the world. Some companies go further than BAe, indicating also the distribution of their assets, by segment and by location. Others indicate the countries in which their products are manufactured, as well as where they are sold.

Most of the data in reports are provided for the preceding year as well as the 'current' one, but for simplicity the extracts in this book usually present only current-year values. Table 3.9 shows the distribution of sales for 1989, the preceding year, as well as 1990. We can see where BAe's sales have risen over the year, and where they

have fallen. However, the sales volumes of aircraft companies are 'lumpy', being much affected by the passage of large orders through the system. Hence great significance cannot be ascribed to the year-to-year change in any one part of the world. Certainly the total volume of sales appears to have risen healthily.

Note 6: *Employees*. British Aerospace gives the total costs of the salaries and other benefits of its employees, and the distribution of its employees among its six business segments. Unlike many companies, BAe does not provide separate information for those employed in the UK and overseas.

Notes 13 and 14: *Fixed Assets*. These give additional detail about the way in which assets such as land and buildings have been valued. This information might suggest, for example, that the balance-sheet values of the assets are significantly lower than their market values. One can also glean information about the ownership of associated companies and other investments.

Note 21: *Contingent Liabilities*. This is the place to look for impending legal actions or the possible need for a write-down in the value of a large investment. British Aerospace's note contains no such worrying suggestions.

Note 28: *Pension Schemes*. The topic of this final note seems placid enough*. However, it could suggest that the company's liabilities have not been adequately recognised in the balance sheet. Liabilities for pension payments — and in some countries for health-care payments for present and retired staff — can have significant consequences for the health of a company. Most BAe employees work in the United Kingdom, where health-care commitments have not escalated severely, and this note contains nothing to concern us.

---

* These words were written in the summer of 1991. Since then the experience of some pensioners — not former employees of British Aerospace or its subsidiaries — has cast doubt upon the proposition. The pension schemes of a few large companies were found to have been eroded by massive misappropriation by the 'trustees' responsible for them.

Non-accountants probably do little more than skim through the notes appended to a company's accounts. This survey indicates the wide variety of information that they are likely to miss, relating to both the company's finances and its operations.

4

Profit and Loss

A conventional Profit and Loss Statement is a superposition of three elements:

● the Trading Account, which shows how the Gross Profit is derived from the Turnover or Sales

- the core Profit and Loss Account, which shows how the company's 'true' Earnings for the year are derived from the Gross Profit, and
- the Appropriation Account, which shows how the Earnings are apportioned to the company's reserves and to its owners, the shareholders.

On the page headed Profit and Loss Account (or Statement) the three accounts are usually stacked one on top of the other. We shall consider them separately, to make explicit the several functions of the complete statement.

4.1 The Trading Account

This is usually presented in a very compressed form, often even shorter than the format shown in Table 4.1. As usual, bracketed quantities are to be subtracted. To understand the significance of this compact statement, we need to look carefully at its components.

Consider first the group's *Turnover*. As first stated, this includes all of the sales of subsidiary companies, plus appropriate fractions of the sales of related or associated companies, each fraction determined by the 'equity' held by companies of the group. The line *Turnover of Related Companies* then subtracts their contributions to obtain the sales of the group's core elements. Sales are determined on an Accruals basis, as explained in Section 2.2.

No comment is required on the deduction of Customs and Excise Duties. They are merely passed on to the collectors of such duties, and play no real part in the company's finances, save by raising prices and thus reducing sales.

Cost of Sales. This quantity is more complex than its name suggests. In typical accounts the amount deducted from Turnover to obtain Gross Profit is the sum of:

> Materials, components and services purchased
> Direct production labour costs
> Depreciation of plant
> Opening stocks

less   Closing Stocks

Only costs relating to the group's production units are included here; other expenses are introduced in the Core Account to be considered later. Hence *Gross Profit* is that generated by the Sales of the group's manufacturing units, before the

Table 4.1   Structure of Trading Account

Turnover or Sales or Revenues
  (Cost of Sales)
  (Customs and Excise Duties)
   Turnover of Related/Associated Companies)
Gross Profit

---

deduction of varied 'overheads' and the possible addition of income from other aspects of the company's operations. The Accruals concept (see Section 2.2) is applied in determining all of these costs.

Depreciation is a complex matter; it will be considered in detail in Section 4.4.

Stocks. This heading gives the total value assigned to partly finished work and to items purchased for resale. The alternative name *Inventory* is used in other parts of the world. In the case of engineering companies, resale normally takes place after substantial value has been added by conversion into more complex or more refined products. Hence Stocks can be broken down into:

- Raw Materials and Components that have not yet been processed
- Work in Progress, to which some value has been added, and
- Finished Goods awaiting delivery.

We now know what Stocks are, but why does their value appear in the Cost of Sales? The accounting problem to be addressed is the following. A company buys large quantities of materials and components during the year, but does not complete the manufacture of products from them. It not only invests in the raw materials, but puts further value into the semi-finished products. Without the correction provided by deducting Closing Stocks from the Cost of Sales, the company might appear to be operating at a loss, whereas in reality it is only the length of the conversion cycle that gives this impression.

The conventional calculation of Cost of Sales treats Opening Stocks as though they were purchased from the preceding year's accounts, and Closing Stocks as though they are sold to the following year's accounts. In this way the trading account seeks to provide a picture of a single year's trading that is less dependent on the timing of the asset-conversion process. These corrections work quite well in 'normal' trading conditions, but show signs of strain when inflation is high or commodity prices fluctuate rapidly, and when applied to companies that are either growing or shrinking rapidly.

Example 4.1 United Scientific Holdings PLC 1990: Valuation of Stocks

---

This company is active in the defence and aerospace sectors, notably through Alvis, a manufacturer of armoured vehicles, and the electro-optical companies Avimo and Helio. The following extract from the Accounting Policies adopted in 1990 illustrates the considerable technical difficulties of valuing stocks.

"Stocks are valued at the lower of cost and net realisable value, less any foreseeable losses and progress payments received on account. Cost comprises prime costs of direct labour and materials together with attributable production overheads.

"Major contracts of a duration in excess of one year have been classified as long term contracts. A prudent assessment of work completed on major long term contracts, valued at cost plus attributable profit, has been included in debtors.

"Progress payments received in excess of the work in progress value of the related contract are included with advance receipts under creditors."

This statement deals explitly with progress payments, payments in advance, and long-term contracts. However, it omits one important rule of valuation, namely, the first-in/first-out (FIFO) convention, the assumption that stocks acquired first are used first. An alternative convention, last-in/first-out (LIFO), is more usual in the United States.

A Note to the accounts gives details of opening and closing levels of USH's stocks.

In the nature of this company's business, there is a significant amount of work in progress, and less than half of its value is covered by progress payments. Note that the overall effect of the year-to-year changes in Stocks is an increase in Cost of Sales, and a reduction in Gross Profit, of some £5.7 million. This is not insignificant, as the reported Gross Profit was £23.7 million.

Table 4.2 Composition of Stocks at Year End (values in £ million)

	1990	1989
Raw materials and consumables	14.7	14.5
Work in progress	29.1	29.3
Long term contract balances	11.5	13.2
Finished goods	2.9	3.3
Progress payments on account	(13.7)	(10.3)
Net Value of Stocks	44.4	50.1

4.2 The Core Profit and Loss Statement

This leads from Gross Profit to Earnings, and then to the Profit attributable to holders of its Ordinary Shares, who are the 'members' or 'owners' of the company. Table 4.3 indicates the significant elements that may appear here. The lines marked (*) will, in some accounts, be placed elsewhere in the stack.

A striking feature of the array in Table 4.3 is the appearance of nine different measures of profit. The statements provided by individual companies do not usually identify all of these stages in the calculation of the final profit, but each item appears in some sets of accounts. Because the calculation is usually much compressed, what is called Trading Profit or Trading Surplus in one set of accounts may elsewhere be called Operating Profit or Profit on Ordinary Activities.

Let us work our way down the stack, to see what the various levels of profit mean. *Trading Profit* is derived from Gross Profit by deducting a number of *Overheads*, necessary costs of running the business and providing for its future. At this stage *Other Operating Income* may be fed into the stream; this might be dividends received from companies classed as Trade Investments.

To get the *Operating Profit*, we add income from certain sources outside the core of the group, notably, equity-defined proportions of the profits of associated companies.

Interest paid and interest received are introduced as we pass to *Profit on Ordinary Activities*; they can affect the profit stream profoundly. Property profits are intro-

Table 4.3   Structure of P & L Statement

Gross Profit
   (Administration costs)
   (Distribution costs)
   (Depreciation of related facilities)
   (*)(R & D and launch costs)
   Other operating income
Trading Profit
   (*)Profits from related/associated
companies
   Licensing income
   Technical fees
Operating Profit
   (Interest expense)
   (*)Interest income
   Investment income
   (Amounts written off investments)
   (*)Property profits
Profit on Ordinary Activities
   (*)(Employees' profit share)
   (*)Exceptional items (+/−)
Profit on Ordinary Activities before Tax
   (Corporation Tax)
   Adjustments (+/−)
Profit on Ordinary Activities after Tax
   (Minority share of profits)
Earnings or Profit on the Year
   Extraordinary items (+/−)
   (*)Currency translation items (+/−)
Profit attributable to All Shareholders
   (*)(Preference dividends)
Profit attributable to Ordinary Shareholders

---

duced here, but a company whose business is trading in property does not put such profits here, but well up the stack, since they come from its basic business.

At this stage allowance may also be made for investments that have gone sour or for gains in selling property. The words *Written Off* indicate that a corresponding deduction is simultaneously made to an asset value recorded on the balance sheet. The entry in the profit and loss statement introduces the reduction on the liabilities side that is necessary to maintain the required balance of assets and liabilities.

If part of the employees' remuneration is related to the company's profits, it can be reckoned only when those profits have been determined. This profit share is deducted as we move to *Profit on Ordinary Activities before Tax*. The *Exceptional Items* are major and infrequent costs (or gains) that relate to the ordinary business of the group. Redundancy payments or less frightening 'restructuring costs' come here, if the businesses undergoing contraction are part of the group's main activities.

The deduction of *Taxation* in moving to *Profit on Ordinary Activities after Tax* seems straightforward, but this single number is derived from tortuous calculations of taxes actually paid and potentially payable to governments around the world. Chapter 8 looks at taxation in more detail. The *Adjustments* may arise from changes in accounting conventions from year to year, or from the need to correct previous errors. Even accountants make mistakes.

In Example 3.2 we recognised the need to assign a fraction of the group profits to the minority shareholders in the group's subsidiaries. This deduction brings us to the group's *Earnings*, sometimes referred to as 'the Bottom Line'. We can see that it is actually not the bottom line. Before reaching the *Profit attributable to Shareholders* we may encounter *Currency Translation Effects*. These are introduced to ensure that the Balance Sheet keeps step with the changing values of the world's currencies.

What are *Extraordinary Items*, and how do they differ from the rather similarly named Exceptional Items? They are distinguished by the nature of their relationship to the ordinary activities of the group. Both are 'material' (that is, of significance to the group as a whole), and infrequent (it is to be hoped, since they are often bad news). However, Extraordinary Items are deemed to fall outside the range of ordinary activities. They might be costs of closing a subsidiary, a loss on its sale at a price below the assigned asset value, or a gain from a sale at a higher price than expected.

We are still not at the end of the trail. Some companies issue *Preference Shares* that entitle holders to a fixed dividend payable (save in the most extreme circumstances) even if the profit does not justify a dividend to the holders of *Ordinary Shares*. Deducting the *Preference Dividends*, if any, we obtain finally the *Profit attributable to Ordinary Shareholders*, the 'members' of the company.

It may have been noticed that profits and income from different kinds of investments are entered at quite different places in the profit and loss statement of Table 4.2. In Section 4.5 further consideration will be given to the ways in which contributions from investments are introduced into group accounts.

4.3 The Appropriation Account

The Core Account looks at profit from the point of view of the shareholders. The Appropriation Account, shown in Table 4.4, adopts the point of view of the company, and indicates how the year's gain has been divided between the shareholders and various internal accounts.

Having set out this part of the Profit and Loss Account, we defer the discussion of most of its features until the next chapter, which deals with the Balance Sheet. That is a more natural place to explain what is meant by 'Share Distribution', 'Goodwill' and 'Reserves'.

Table 4.4 Structure of Appropriation Account

Profit attributable to Ordinary Shareholders
    (Ordinary dividends)
    (Share distribution)
    (Goodwill written off)
    *Currency translation items ($+/-$)
    Transfers to Reserves ($+/-$)
Profit retained in Profit and Loss Account

Earnings per Share. At the very bottom of the page headed Group Profit and Loss Account, following the appropriation section, one usually finds the item *Earnings per Share*, widely known by the abbreviation 'EPS'. This quantity is obtained by dividing the Earnings by the number of ordinary shares in issue, usually the average number over the year. There is great interest in this quantity, since it is a measure of the gain of the ordinary shareholder from the year's trading.

Although the concept of Earnings per Share is simple, technical problems arise in evaluating both numerator and denominator. They are dealt with in Sections 6.1 and 9.9.

The distinction between 'exceptional' and 'extraordinary' items becomes important here. The former are said to be 'above the line' (that is, they affect the Earnings), while the latter are 'below the line' (that is, below the line at which the Earnings are struck). Moreover, exceptionals may affect the corporation tax that is due, since they come above the point at which the Taxable Income is determined. Hence the directors' decision to classify an expense (or a profit) in one of these two ways can have important effects on both tax and earnings per share. Ironically, the choice that keeps EPS from falling after an unusual loss may have the effect of increasing the tax bill. What appears superficially attractive to the investor may cost the company money. In practice, Her Majesty's Inspectors of Taxes can be relied upon to ensure that companies are unable to avoid taxation without good reason.

Example 4.2 PowerGen plc 1991: Consolidated Profit and Loss Account

---

The accounts set out in Table 4.5 were the first issued by this company after privatisation early in 1991. It operates electricity-generating plants throughout England and Wales, supplying some 28 per cent of the country's needs through a 'pool' operated by The National Grid Company.

Note that PowerGen's statement moves directly to Operating Profit. Even the Notes do not provide all of the information needed to calculate the Gross Profit and the Trading Profit.

PowerGen shows both Exceptional and Extraordinary costs. The Exceptionals recognise equalisation of employees' retirement ages, and provisions for liability and damage claims. The Extraordinaries recognise other liabilities and costs of privatisation. The need to equalise retirement ages arose from a European Court ruling, and illustrates the growing influence that European institutions have on the finances of British companies.

Table 4.5   Consolidated Profit and Loss Account to 31st March 1991 (values in £ million)

Turnover	2651.2
Operating costs	(2384.7)
Operating profit before exceptional items	266.5
Exceptional items	(25.6)
Operating profit	240.9
Net interest receivable	59.8
Profit on ordinary activities before taxation	300.7
Tax on profit on ordinary activities	(102.2)
Profit on ordinary activities after taxation	198.5
Extraordinary items	(54.4)
Profit for the year	144.1
Dividends	(43.4)
Retained profit for the year	100.7

This set of accounts is not entirely representative of those that will be issued in years to come, since the interest payable was unusually low prior to privatisation. PowerGen has also provided 'pro forma' accounts, which seek to show how the company would have performed, if it had carried higher debt throughout the financial year. They show 'restated' Earnings of £179.7 million and Profit for the Year of £125.3 million.

Two values have been quoted for Earnings per Share: 23.0 p (derived from the Earnings) and 25.2 p (excluding exceptional items). In neither case have the 'actual' Earnings been used in the calculation; the company believes that the 'restated' ones give a better representation of the year's activities. Thus

£179.7 million/(781.3 million shares) gives £0.23 per share.

4.4 Depreciation

What is It?

In everyday use, the word Depreciation means 'loss in value over a period of time'. This is not what it means in accounting, where it is seen as a way of spreading major expenditures over a number of years, to prevent the accounts from jumping about from year to year in a startling and unhelpful manner. To recognise the need for this accounting device, we must imagine what the accounts would look like if it were not used. This will be done in the example which follows.

For any one year the total depreciation charge is the sum of numerous contributions, each relating to a different asset purchased at some point in the past. Depreciation 'expense' is not deposited in a special account that can ultimately be drawn upon to purchase replacements for the depreciated assets. The 'funds' associated with this notional expense are actually deployed in varied ways during the financial year for which the charge is made. Another word for Depreciation is *Amortisation*, but this is commonly used in a special way which will be introduced in Section 5.6.

Example 4.3 Illustrating the Role of Depreciation

Consider an imaginary company which initially has Sales of 1000, giving a Gross Profit of 300 after Cost of Sales of 700. Suppose that this company invests 1000 in a major expansion, which doubles both Sales and Cost of Sales (the latter excluding the investment itself). Accounting for capital expenditure as it occurs, the company would produce the profit and loss account of Table 4.6. For simplicity, no allowance is made for tax,

for growth of the business or for inflation. It is necessary to deduct interest on the invested funds (a rate of 15 per cent has been assumed), so the investment does not appear to be unrealistically beneficial.

Two points can be made about the progression of annual profits shown in Table 4.6. They fluctuate wildly as the investment is made. However, the augmented earnings will, after a number of years, compensate for

Table 4.6   Profit and Loss Account without Depreciation

	Year 0	Year 1	Year 2	Year 3	Year 4
Sales	1000	2000	2000	2000	2000
Costs of Sales	(700)	(1400)	(1400)	(1400)	(1400)
Investment	–	(1000)	–	–	–
Gross Profit	300	(400)	600	600	600
Interest	0	(150)	(150)	(150)	(150)
Profit	300	(550)	450	450	450

Table 4.7   Profit and Loss Account with Depreciation

	Year 0	Year 1	Year 2	Year 3	Year 4
Sales	1000	2000	2000	2000	2000
Costs of Sales	(700)	(1400)	(1400)	(1400)	(1400)
Depreciation	–	(100)	(100)	(100)	(100)
Gross Profit	300	(500)	500	500	500
Interest	0	(150)	(150)	(150)	(150)
Profit	300	350	350	350	350

the apparent loss in the year of investment, so that the profit averaged over a number of years is relatively stable.

The introduction of depreciation, as shown in Table 4.7, spreads the cost of investment over a number of years, thus limiting the unevenness of the profit stream, and giving a better indication, year by year, of the long-term profitability of the company. It has been supposed that the cost of the investment is spread uniformly over ten years.

The particular assumptions made in creating this example give rise to a slight increase in profit following the major invest-ment. There is no fundamental reason why this should be so: a longer depreciation period or a lower interest rate would give rise to higher profits; a lower sales volume would reduce them.

Although the device of depreciation has 'sheltered' the profit and loss statement from the impact of the major outlay, such invest-ment activities will still affect the balance sheet, the funds flow statement, and cash flow statement, if one is provided. This is illustrated by the purchase of Jaguar plc by Ford UK, whose effects were noted in Example 3.2.

How is Depreciation Calculated?

The most straightforward method of calculation is *Straight-Line Depreciation*, which spreads the cost uniformly over the number of years over which an asset is expected to be useful, with allowance for its *Scrap Value* at the end of the period. Thus

$$\text{Annual Depreciation} = \frac{\text{Cost} - \text{Scrap Value}}{\text{Assumed Useful Life}} \qquad (4.1)$$

Another much-used procedure is *Reducing-Balance Depreciation*, in which a constant fraction of the residual value is written off annually. This has the effect of introducing a larger depreciation charge in the early years.

No matter how it is arrived at, the value appearing on the balance sheet at any point in time is referred to as the *Book Value*.

The 'useful life' that is chosen depends on the nature of the asset considered, as is indicated in the following example. The 'life' selected may turn out to be unrealistically long, possibly because a newly constructed plant fails to operate as planned, or because the anticipated market for its products does not materialise. In such circumstances an Exceptional or Extraordinary charge may be required to write off the residual, undepreciated value.

On the other hand, an asset may remain in use long after it has been fully depreciated, leaving a negligible book value on the balance sheet. It has recently been proposed, for example, that some of the Magnox class of nuclear power stations should remain in operation for forty-five years, rather than the thirty years envisaged when they were constructed.

The inclusion of depreciation expenses in the accounts for a particular year has important implications. Although they are claimed to describe the financial transactions of a past year, annual statements are significantly influenced by judgements about what is likely to happen in the years to come. Depreciation is not the only means by which expectations of the future modify a year's accounts. The inclusion of opening and closing stocks (Section 4.1) and provisions (Section 5.3) also involves judgements about future events and their effects on the values of assets and liabilities.

Example 4.4 PowerGen plc 1991: Depreciation of Assets

This company's profit and loss statement was considered in Example 4.2. Buried within the cost of sales are two elements of depreciation:

Depreciation of owned assets £75.0 million

Depreciation of leased assets £2.7 million

To see how these numbers are obtained, we look first to the Accounting Policies:

"Depreciation of Tangible Fixed Assets. Provision for depreciation is made so as to write off, on a straight line basis, the book value of tangible fixed assets including those held under finance leases. Assets are depreciated from the dates they are brought into use over their estimated useful lives or, in the case of leased assets, over the lease term if shorter. No depreciation is provided on freehold land. The estimated useful lives for the other principal categories of fixed asset are:

Power stations	25 to 40 years
Non-operational buildings	40 years
Short term assets	5 years"

Short-term assets include, for example, motor vehicles and computing equipment. Some companies identify different useful lives for many more categories of assets.

Details of the calculation of depreciation are given in the Notes to PowerGen's accounts. By way of illustration, we consider in Table 4.8 two classes of assets, Generating Plant and Assets in Course of Construction. No depreciation has been charged to Assets under Construction, as explained in the Accounting Policies.

The current book values of these two classes of assets are calculated in Table 4.9. Here we begin with the acquisition cost and deduct from it the total depreciation that has accumulated since acquisition. Alternatively, we can proceed as in Table 4.10, starting from the preceding year's book value. The published information does not tell us how the depreciation is distributed between the individual assets of the group.

Table 4.8   Depreciation of Plant (values in £ million)

	Generating Plant	Assets under Construction
Depreciation to 1 April 90	515.7	–
Charge for the year	47.2	–
Effects of asset disposals	(2.2)	–
Depreciation at 31 March 91	560.7	

Table 4.9   Calculation of Book Values (values in £ million)

	Generating Plant	Assets under Construction
Cost of assets 1 April 90	1481.3	90.2
Additions during the year	56.6	110.1
Disposals during the year	(30.2)	–
Depreciation at 31 March 91	(560.7)	–
Book Value at 31 March 1991	947.0	200.3

Table 4.10   Changes in Book Values (values in £ million)

	Generating Plant	Assets under Construction
Book Value at 31 March 90	965.6	90.2
Additions during the year	56.6	110.1
Disposals during the year (net)	(28.0)	–
Depreciation for the year	(47.2)	–
Book Value at 31 March 91	947.0	200.3

4.5 Income from Investments

The profits generated by subsidiary companies, related companies and other investments are shown in accounts in quite different ways, to the despair of students of finance. The procedures adopted are implicit in the Profit and Loss Account set out in Section 4.2, but they may not be appreciated without further explanation.

Subsidiary Companies. Whether subsidiaries are wholly owned or not, their complete sales, costs of sales and profits are shown in the group accounts. It is only when the Minority Interests line is reached that the profit and loss statement acknowledges the fraction of the profits attributable to the minority shareholders.

Related or Associated Companies. The Group Turnover includes equity-defined proportions (normally between 20 and 50 per cent) of the sales of related or associated companies, but this element is deducted, as is the Cost of Sales, in determining the Gross Profit. Later the appropriate fractions of the operating profits of associated companies are added to the operating profits of subsidiaries to determine the group Profit on Ordinary Activities.

Trade Investments. The dividends (not the profits) received by group companies in recognition of their holdings of shares below the 20 per cent threshold are entered as Other Operating Income in determining the Trading Profit of the group. Since many companies do not distinguish between Gross Profit and Trading Profit, this dividend income may not appear on the profit and loss statement, being identified (if at all) only in a Note. It then acts as a negative contribution to the Cost of Sales.

Interest Received. Investment need not take the form of shareholdings, but may be in the form of interest-bearing securities. The resulting interest income may be introduced at the operating level (thus contributing to Operating Profit) or may be combined with Interest Expense to provide a single Net Interest figure, which is entered after the Operating Profit has been found.

Acquisitions and Divestments. When companies join or leave a group, their contributions to turnover and profits are normally included in the group accounts only for the periods during which they were part of the group.

Example 4.5: Idealised Profit and Loss Statements

To illustrate the ways in which profits and income from different kinds of investments are entered into a group's accounts, we shall consider a number of hypothetical organisations. Each uses the same amount of capital, but it is invested in the different ways set out below. To simplify the comparisons, we assume that all of the enterprises are equally profitable, with a 20 per cent margin on sales. The fraction of the profits paid as dividends will also be taken to be uniform (at one half) for all the companies involved. Again to simplify, we omit complications such as tax and interest expenditure.

Case A: All the capital in wholly owned subsidiaries. The group accounts are then essentially the same as those from a single company without subsidiaries.

Case B: Half the capital in wholly owned subsidiaries, and the other half used to purchase just over 50 per cent of the shares in further subsidiaries. Here we must allow for the interests of the minority shareholders in the part-owned subsidiaries.

Case C: Half the capital in wholly owned subsidiaries, and the other half used to purchase just over 20 per cent of the shares in a number of associated companies. Here we must adjust the group turnover and cost of sales, as well as introducing a new profit stream.

Case D: Half the capital in wholly owned subsidiaries, and the other half purchasing 10 per cent interests in a number of trade investments. The group turnover will not reflect the activity within these investments, and only dividends received from them will appear as group profit.

Case E: Half the capital in wholly owned subsidiaries, and the other half in interest-bearing securities. The interest rate will be taken to be 16 per cent.

The P & L statements derived from Tables 4.1, 4.3 and 4.4 are set out in the Table 4.11. Towards the bottom are indicated payments of dividends and the funds retained by the companies considered. It will be realised that the accounts of most groups combine all of the features that have been presented here in isolation.

A number of principles of business strategy are illustrated by this tabulation:

- Each of the investments in companies (*Cases A* to *D*) generates the same dividend payments to the holding company, on the plausible assumptions adopted.
- Part-owned subsidiaries bulk up a group's Turnover, and enhance the funds under its control.
- Control of a company is valuable; lacking it, the parent company does not actually

Table 4.11 Idealised Profit & Loss Statements

	Case A	Case B	Case C	Case D	Case E
Group Turnover	1000	1500	1000	500	500
Turnover of Associates	–	–	(500)	–	–
Reduced Group Turnover	1000	1500	500	500	500
Cost of Sales	(800)	(1200)	(400)	(400)	(400)
Gross Profit	200	300	100	100	100
Dividends from Investments	–	–	–	50	–
Trading Profit	200	300	100	150	100
Profits of Associates	–	–	100	–	–
Operating Profit	200	300	200	150	100
Interest Income	–	–	–	–	80
Profits on Ordinary Accounts	200	300	200	150	180
Minority Interests	–	(100)	–	–	–
Earnings	200	200	200	150	180
Dividends	(100)	(100)	(100)	(75)	(90)
Profit Retained	100	100	100	75	90
Other Features:					
Dividends received	50	50	50	50	–
Dividends to Other Shareholders	–	50	200	450	–
Retained in Other Companies	–	50	250	500	–
Retained controlled by Group	100	150	50	75	90

have access to some of the Earnings reported in its accounts.

- Sub-associate shareholdings make a reduced contribution at the 'bottom line', as well as offering the parent company little influence over the use of the profits generated.

4.6 Accounting for Inflation

Historical Cost Accounting

According to this convention, assets retain their cost at acquisition or manufacture, subject to depreciation where appropriate. This approach displays Objectivity, since the records of purchase and manufacture (or past accounts) can be examined to determine values. Moreover, the accounts are simpler, since adjustments need not be made to allow for changing market values of the assets.

In times of rapid inflation this convention gives rise to book values that become progressively smaller fractions of market values, and of current replacement values of depreciating assets. Some effects on accounts produced in times of high inflation will now be identified.

Consider first the consequences of inflation for Depreciating Assets. Depreciation based upon acquisition value ceases to represent the intended fraction of the present-day cost. Thus it fails to perform its intended function of spreading the real cost over a number of years. By making too small a contribution to the Cost of Sales, depreciation then gives rise to an apparent profit which is unrealistically high. This artificial profit is subject to taxation, which takes funds from the company which should be retained to sustain its future operations.

Turning now to Stocks, we note that the use of the first-in/first-out valuation convention means that the same physical quantity of stocks has a higher money value at the end of the year. Thus Closing Stocks will exceed Opening Stocks, further reducing the Cost of Sales and further increasing the apparent profit and its exposure to taxation.

Two other consequences of the application of historic-cost methods in times of high inflation may be noted. The generation of apparently high profits often leads to a corresponding large dividend pay-out, drawing from the company funds that may be needed for its continued health. Finally, as recorded asset values fall behind realistic market values, the balance sheet fails to provide even an approximate valuation of the company.

Current Cost Accounting

This is one of the schemes devised to overcome the difficulties engendered by inflation. It was quite widely used in the United Kingdom in the 1970s and 80s, when inflation was high, but has generally been abandoned as inflation rates have fallen back. Four main corrections are applied to the conventional, historic-cost accounts:

- a Cost of Sales Adjustment, intended to base the cost of stocks on the time of consumption rather than the time of purchase
- a Monetary Working Capital Adjustment, allowing for the need to keep a larger volume of liquid assets in the company, in addition to a larger monetary value of stocks
- a Depreciation Adjustment, achieved by periodically revaluing assets at their market or replacement values, with the annual depreciation charge (see Equation (4.1)) based on these enhanced values
- a Gearing Adjustment, which reduces the effects of the preceding corrections in proportion to the capital invested by lenders, rather than by shareholders.

Accounting for inflation is made more complicated by the existence of different rates of inflation in the several countries in which a group does business. The accounts of each company of the group must be adjusted for the inflation it has experienced. These sets of accounts are then superimposed to give the group accounts, and corrections with labels such as 'Currency Translation Effects' are introduced to tidy things up.

PLATE 4

The 'Thames Bubbler' operating with oxygen supplied by the BOC Group.

Example 4.6 The BOC Group plc 1982: Profit and Loss
for Various Accounting Conventions

This group is a major worldwide supplier of industrial gases, and is also engaged in health care (pharmaceuticals and anaesthesia equipment) and other activities, notably, high-vacuum technology. In 1982, during a period of high inflation, it produced four distinct profit and loss statements:

- a conventional historical cost statement
- a 'modified historical' statement, with limited corrections that the company felt were appropriate
- a current cost statement of standard form
- an income statement according to US Generally Accepted Accounting Practices (GAAP, historical cost).

This last statement is included in the Table 4.12 to demonstrate the sensitivity of the profit calculation to the conventions adopted in preparing the accounts. The other statements make the same point, but for different reasons.

The arrows on the right-hand side of this table mark the incidence of corrections and differing conventions. As might be expected, the conventional accounts give rise to a significantly larger 'Profit' than do the other three statements. It is something of a coincidence that the US GAAP statement produces much the same profit as do the inflation-adjusted accounts. Note that it is the different treatment of Taxation that reduces the profit when the US conventions are applied. Moreover, even the Turnover is calculated in a different fashion.

The Transfers to Reserves (or retained profits) given by the four calculations are quite different. The historic-cost statement suggests substantial retentions, while the others indicate retentions little more than half as large.

Table 4.12   Results Presented by the BOC Group in 1982

	Historical Cost	Modified Historical	Current Cost	US GAAP	
Turnover	1534	1534	1534	1476	←
Cost of Sales	(861)	(920)	(920)	(827)	←
Gross Profit	673	614	614	649	
Operating Profit	203	143	138	198	
Realised Stock Holding Gains	–	18	–	–	←
Trading Profit	203	161	138	198	
Gearing Adjustment	–	–	26	–	←
Interest (net)	(58)	(65)	(65)	(58)	
Interest Capitalised	–	7	7	–	←
Profit before Tax	145	103	106	140	
Tax	(28)	(28)	(28)	(58)	
Minority Interest	(13)	(11)	(11)	(13)	
Profits from Discontinued Operations	–	–	–	5	←
Earnings	104	65	67	74	
Extraordinary items	6	6	6	–	←
Profit	110	71	73	74	
Dividends	(20)	(20)	(20)	(20)	
Transfer to Reserves	90	50	53	54	
Return on Equity (Earnings/SHF)	20.8%	7.7%	8.0%	14.9%	

The adoption of a different accounting convention will affect not only the profit and loss account, but the balance sheet. The specific effects on the shareholders' funds are not considered here, but the opportunity is taken to indicate (at the bottom of the table) the Return upon Equity given by the four conventions. The returns from the inflation-corrected accounts are likely to be better indicators of the real earnings of the company and its shareholders. As might be expected, they are much more modest than the return indicated by the conventional calculation.

4.7 Some Words of Warning

Before we leave the profit and loss statement, it is worth noting some of the ways in which it can mislead. We have already identified some of the perils of inflation: unrealistically low book values for assets, and unrealistically high profit levels largely reflecting inflation rather than real retentions. We noted also that directors have considerable discretion in deciding whether unusual costs are declared as 'exceptional' (and thus affect the Earnings) or as 'extraordinary' (and therefore do not). Indeed, they must decide when the probability of an irremediable fall in value is so great that a write-down in valuation is appropriate.

The figure for Sales or Turnover does not always represent hard sales for which payment has been received or ever will be. The goods may still be in the shipping room; the prices assigned may (or may not) include discounts, freight costs, insurance and sales tax; and allowance may not have been made for the possible return of merchandise.

A rapid acceleration in Sales is not always a good sign. The growth rate that can be sustained by the company's trading may be exceeded, so that it will have to seek more capital, either by borrowing or by issuing more shares. Commonly, this is a situation in which a healthy profit is stated, while the company remains chronically short of cash. These are circumstances in which a company is particularly vulnerable to adverse changes in the trading environment, either general or specific to its markets.

There are a number of ways in which stated expenses can be reduced temporarily. One year's results are sometimes improved by bringing forward Tax Losses arising from earlier years of unprofitable trading. Another temporary reduction in expenses is a 'pensions holiday', during which the company does not contribute to a pension fund whose assets exceed the anticipated calls upon it. This may be the result of unexpected growth in the pension fund's investments, or may be the consequence of high turnover of staff or reduction in the number of staff employed. Sometimes it even proves possible to withdraw cash from the pension fund, and this appears as income in the accounts. (In 1991 the Occupational Pensions Board agreed that £150 million of surplus contributions might be refunded to Lucas Industries plc; £60 million of that amount went to Inland Revenue, and the company planned to use the rest to reduce its borrowings.)

The periods chosen for depreciation or amortisation of assets are little better than guess-work. Also, the depreciation charge for the year depends on the point at which the depreciation of an asset is initiated, that is, when it is taken on to the books, rather than assumed to be in the process of construction or acquisition. Note too that interest on loans taken out to fund the construction of a major asset may be *capitalised* during the period of construction. This means that the interest is added to the asset's acquisition value, rather than being charged as an expense as it is paid. Thus the interest appears on the balance sheet, but not in the profit and loss statement.

Income claimed for a year may include dividends declared, but not yet paid; and the profit shares from associated companies will not have been remitted in cash, save for the dividends that have been declared and paid. The profits of overseas elements of a group may not be repatriable; even cash, the ultimate 'liquid asset', can be trapped abroad. It is very difficult to represent meaningfully the affairs of companies operating in regimes of very high inflation. In such circumstances the measures adopted to 'manage' currencies can have important effects on the group's finances, not all of which may be acknowledged in the published accounts.

The income statement applies to an entire group. Even when overall health is indicated, particular components of the group may be trading at a loss. The nature of inter-group and inter-segment transfers is seldom made clear.

These warnings are not intended to imply that accounts are commonly prepared with the intention to deceive – though some are. Rather, they remind the reader of the complexity of the organisations and the activities that the compact financial statements seek to describe. Even the financial officers of a company may fail to understand its true financial position until it is too late for effective corrective action.

An analogy may help the reader to understand and accept the various artifices adopted in generating the accounts for a particular trading year. In reply to a query about the direction in which a sailing-boat is moving, the bearing of its intended destination may be more helpful than the heading which variable winds have forced its crew to adopt temporarily.

5

Assets and Liabilities

The balance sheet is the foundation of accountancy. Until fairly recent times, British companies were required to provide only this single financial statement in their annual report to shareholders. Given very comprehensive notes on the balance-sheet entries, and the algebraic results of Section 3.2 (Equations 3.1 to 3.3), one could construct the other now-standard financial statements. Even though this process is not necessary, the balance sheet remains the core of a company's accounts.

The concepts behind the balance sheet were developed about the time that Columbus made his celebrated Atlantic voyages. At that time an Italian monk named Paccioli produced a book on algebra, with an appendix on double-entry bookkeeping, the basis of the balance sheet. The values entered on a company's balance sheet are the current totals extracted from a series of ledgers maintained by the company's staff. Each ledger is driven by the practices of double-entry bookkeeping, and the day on which they are ruled off is the *Balance-Sheet Date*.

Although a balance sheet is not intended to be a valuation statement, it has the appearance of one. Just after a company is established, when its balance sheet is first drawn up, the assets and liabilities have 'real' values: they approximate to the actual capital invested and the market values of the assets. Section 4.6 introduced one of the progressive mechanisms that drive balance-sheet values away from market values: currency inflation. A second mechanism, which introduces impulsive change rather than the insidious drift of inflation, is the adjustment required when subsidiary companies are added to the group. The merging of their balance sheets into that of the parent group often involves a reduction in the apparent capital employed by the group and in the balancing assets. This process will be discussed in Section 5.6.

5.1 Balance-Sheet Formats

More Algebra

The most compact way of representing the alternative layouts of the balance sheet is to extend the algebraic models of Section 3.2. Equation (3.1) expressed the balance of assets and 'liabilities' as

$$A = L + S$$

where L represents the liabilities proper, and S the shareholders' funds, also seen as a liability of the company. In Example 3.2, when considering the assets of the Ford Motor Company, we saw that this statement is incomplete. It does not take account of Minority Interests, the funds invested in subsidiaries by their minority shareholders. Denoting these by the letter M, we generalise the balance to

$$A = L + S + M \tag{5.1}$$

To define the balance-sheet formats in general use, we introduce further symbols to represent the components of Assets and Liabilities that appeared in Example 3.2, in which the balance sheet of the Ford Motor Company was examined:

$$A = A_f + A_c \tag{5.2}$$

Total Assets = Fixed Assets + Current Assets

$$L = L_l + L_c \tag{5.3}$$

Liabilities Proper = Long-term Liabilities + Current Liabilities

The diagrams of Fig. 5.1 show the three balance-sheet formats to be considered.

Using the symbols introduced above, we obtain a more detailed structure for the balance sheet:

$$A_f + A_c = L_l + L_c + S + M \qquad (5.4)$$

Total Assets = Total Liabilities

This can be described as a *Gross Balance Sheet*, since every quantity in it is given in its total form. In this presentation the assets and liabilities are often placed side by side, rather than one below the other as in Fig. 5.1.

While the format of Equation (5.4) is not now used by British companies, it is widely adopted abroad, by both European and American companies. But there is one significant change in the layout. Usually the Current Assets and Current Liabilities are placed at the top of their lists, since they are closer to the cash which is required day-to-day to run the business. The formats conventional in British accounts are *Net Balance Sheets*. Transposing the Current Liabilities, we obtain

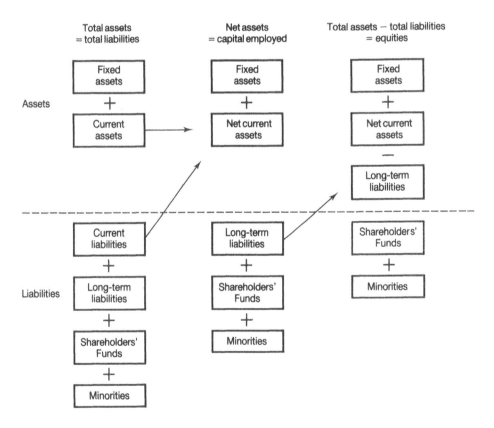

Figure 5.1   Balance sheet formats

$$A_f + (A_c - L_c) = L_c + S + M \qquad (5.5)$$

Fixed Assets + Net Current Assets
= Long-term Liabilities + SHF (including Minorities)

or Net Assets = Capital Employed

This format adopts an *Entity Approach*, by balancing the net assets of the company against its long-term liabilities.

The alternative *Proprietary Approach* balances the shareholders' funds against the other components:

$$A_f + (A_c - L_c) - L_c = S + M \qquad (5.6)$$

Total Assets − Total Liabilities Proper = SHF (incl. Minorities)

This is the presentation now most widely adopted by companies reporting in the United Kingdom.

Example 5.1 Ford Motor Company 1989: Balance-Sheet Formats Compared

---

The gross balance sheet was presented in Example 3.2. Tables 5.1 and 5.2 show the two net formats. The latter is that actually adopted by Ford UK.

Note that the sum of Ford's Net Current Assets was negative at this balance-sheet date. This was not so a year earlier; the indebtedness associated with the purchase of Jaguar plc produced the change. This apparent deficiency in liquid funds is not a matter of great concern for Ford UK, which is a subsidiary of an international group with very large financial resources.

Table 5.1 Balance Sheet in Net, Entity Format

Fixed Assets (A_f)		3298	
Current Assets (A_c)	4514		
Current Liabilities (L_c)	(5578)		
Net Current Assets		(1064)	
Total Assets less Current Liabilities			2234
Long-term Liabilities (L_l)		1262	
Share Capital	39		
Reserves	900		
Shareholders' Funds (S)		939	
Minority Interests (M)		33	
Capital Employed			2234

Table 5.2 Balance Sheet in Net, Proprietary Format

Fixed Assets (A_f)		3298	
Current Assets (A_c)	4514		
Current Liabilities (L_c)	(5578)		
Net Current Assets		(1064)	
Long-term Liabilities (L_l)		(1262)	
Total Assets less Total Liabilities			972
Share Capital	39		
Reserves	900		
Shareholders' Funds (S)		939	
Minority Interests (M)		33	
Capital Employed			972

Another point of interest is the small value assigned to Ford's Fixed Assets, relative to its Current Assets. Within the Accounting Policies appended to the accounts, we note the statement: "Tangible assets are stated at original cost less government grants and depreciation". It appears that the company does not attempt to value its assets in present-day currency values, nor at current market prices.

5.2 Classification of Assets

We now look inside the broad headings of the balance sheet, to see what they include. We start with the two kinds of assets distinguished in Equation (5.2): Fixed and Current.

Fixed Assets

Despite the name, these are not always screwed down. Rather, they are assets that are expected to be of continuing use to the business and are not readily convertible into cash. Three categories are distinguished:

Intangible Assets
Tangible Assets
Investments

They will be considered in turn.

Intangible Assets. This heading does not appear in the accounts issued by most companies. Intangible assets are those without a physical identity, such as patents, copyrights, trade marks, brand names and licences. This list omits some important stores of value within a business: the knowledge and skills of its employees, its product designs and varied methodologies, and established relationships with other companies. Nor does the list mention the value that may be assigned to obtaining effective control over a business. Example 4.5 suggests some of the benefits, in particular, access to the whole of the profits and cash generated by a subsidiary company.

Intangible assets are usually omitted from the balance sheet because their inclusion is thought to conflict with the fundamental accounting concept of Objectivity. On the other hand, it can be argued that the financial position of the company may be grossly misrepresented unless some attempt is made to assess the value of these intangibles. Indeed, it can be argued that their omission from the balance sheet conflicts with the accounting principle that a company should be viewed as a Going Concern. The treatment of intangible assets is a highly contentious issue within the accounting profession.

When a subsidiary is acquired, the purchase price is often much higher than the net value of its identifiable assets. The premium acknowledges the various intangible elements identified above. It is given the name *Goodwill*; ironically, such transfers of ownership often generate much bad feeling. Section 5.6 indicates a number of ways in which this quantity can be entered into the accounts of an acquiring company.

Tangible Assets. Now we come to something more straightforward. These are such things as land, buildings, plant, machinery, motor vehicles, tools and equipment. We have seen that the assigned book values are depreciated (save for land) over what are expected to be their useful lives. Moreover, revaluation may take place as a consequence of inflation or changing trading conditions.

Investments. These are nearly intangible, being mostly in the form of share or loan certificates, issued either by subsidiary and related companies, or by companies outside the group. Only loans more than a year from repayment are shown as Fixed Assets.

Pieces of paper they may be, but many of these assets have readily determined acquisition or market values. They can therefore be entered on the balance sheet with a considerable measure of objectivity. Other investments – principally shares in unlisted companies, that is companies whose shares are not listed by a stock exchange – must be valued in some other way.

Balance Sheet of the Parent Company. The standard package of financial statements includes a balance sheet for the holding company, as well as the consolidated balance sheet for the group it controls. The parent-company balance sheet is not as interesting as that for the entire group – except, presumably, to accountants. As might be expected, the assets of holding companies mostly fall within Fixed-Asset Investments, the entry which subsumes shares in subsidiary companies.

Since the companies of the group are distinct legal entities, the financial collapse of one of them need not affect the others, in particular, the parent company. However, if the parent company has lent money to a subsidiary that fails, that investment may be lost. The total of such loans is shown in the parent-company balance sheet, and indicates the exposure of the parent to this kind of risk.

Current Assets

Here the classification is somewhat wider and includes:

Stocks
Debtors
Investments (again)
Cash at hand or at bank
Fixed assets intended for disposal

Nothing need be said here about *Stocks*, since this item was considered in Section 4.1, when the profit and loss account was examined.

Debtors. The assets labelled Debtors do call for comment. Somewhat confusingly, this word refers to amounts owed to group companies. A clearer alternative name is *Accounts Receivable*. Although this is classed as a 'current' asset, the heading includes all amounts due, both within a year and beyond. Debtors are identified by such labels as:

Trade debtors
Amounts owed by group companies
Amounts owed by related companies
Prepayments and accrued income

The Trade Debtors are perhaps of greatest interest; these are amounts not yet paid by those who have received goods and services from group companies. Obviously, it is a matter for concern if this entry rises significantly without adequate justification.

Current-Asset Investments. These differ from Fixed-Asset Investments in being capable (the company hopes) of rapid conversion into cash. Examples are government securities and shares classed as Trade Investments. Some of these 'financial instruments' are virtually indistinguishable from cash, since they are highly marketable.

Remember that Fixed Assets are those the group intends to use indefinitely. Once the decision has been made to dispose of certain assets, there is a case for including their value under Current Assets. This explains the category *Fixed Assets intended for Disposal*.

Example 5.2 Trafalgar House PLC 1990: Analysis of Assets

The companies controlled by this group are active in a number of industrial sectors, and its assets are correspondingly varied. Some of the group companies are: Trollope & Colls and Cementation (construction), Ideal Homes, Cleveland Structural Engineering, John Brown (power and process plant, plastics processing machinery), Cunard (passenger and container shipping), the Ritz Hotel London, and Heavylift Cargo

Airlines. (In 1991 the Davy Corporation was absorbed, extending the range of Trafalgar House's work in the design and construction of process plant.)

We shall see how the group includes these varied assets in its balance sheet, drawing upon the Notes to the Accounts for the detail that is necessary to obtain any real understanding of what lies behind the bland headings of the balance sheet.

Trafalgar House records no Intangible Assets on its balance sheet. It values its Tangible Fixed Assets in two ways, using the historical-cost convention, and using recent valuations. From Table 5.3 it is evident that the company revalues only its freehold property, hotels and business premises. It is the higher total, based on revaluation, that is shown on the group balance sheet.

The long-term, Fixed-Asset Investments of the group are shown in Table 5.4. The last investment in the list is not recognised on the balance sheet, where only £102.5 million is shown, the sum of the first three items.

The Current Assets identified by Trafalgar House are those of Table 5.5. Two large items call for comment. Under the heading Debtors are indicated large payments expected from customers: accounts falling due within a year (£272.9 million) and accounts recoverable on contracts

Table 5.3 Tangible Fixed Assets (values in £ million)

	Historic Cost	Recent Valuation
Business Premises: Freehold	34.9	39.3
Long Leasehold	16.5	16.5
Short Leasehold	9.4	9.4
Hotels: Freehold	40.5	118.8
Long Leasehold	16.2	16.2
Ships and Aircraft	284.6	284.6
Plant and Equipment	136.1	136.1
	538.2	620.9

Table 5.4 Fixed Asset Investments (values in £ million)

Associated companies: Listed shares	3.7
Unlisted shares	40.3
Loans	58.5
Other: investment in Hardy Oil & Gas plc	20.0

Table 5.5 Current Assets

Developments for sale	703.3
Stocks	80.5
Debtors	681.6
Investments	36.1
Cash at bank and in hand	280.5
	£1782.0 million

(£200.1 million). Developments for Sale represents commercial property in the United Kingdom and United States and residential property there and in Europe. It is reported that £57.3 million of capitalised interest is included under this heading. In view of the slackness in the property market in 1990, one might wonder why this property has been reported as a 'current' asset.

PLATE 5

Construction at Canary Wharf by Trafalgar House subsidiaries.

We note that the group appears to have sufficient cash in hand (some £280 million) to cover its running costs.

An engineer looking at the accounts of Trafalgar House might find the property side of the group's business of little interest. He could be wrong, for financial instability in any part of a group inevitably affects every group company, possibly through a cash famine, possibly through adverse publicity, or possibly through forced sales to restore the financial position. These remarks should not to be interpreted as suggesting that Trafalgar House is, or was, under financial strain. They illustrate the kinds of questions that the informed reader will have in mind while examining a company's balance sheet.

In Example 5.3 the Current Liabilities of Trafalgar House will be analysed. Bringing forward their total, we calculate the Net Current Assets in Table 5.5. Without the inclusion of Developments for Sale among the current assets, the Net Current Assets would be negligible – only (!) £36 million. Indeed, for the prior year (1989, for which values are not given here), the NCA would have been negative without that element of current assets.

Table 5.6   Net Current Assets

Current Assets	£1782.0 million
Current Liabilities	1042.7
Net Current Assets	£739.3 million

5.3 Classification of Liabilities

From the lips of Polonius, Shakespeare brings forth the advice: "Neither a borrower, nor a lender be". This is not generally taken to heart by commercial organisations. Most manufacturing companies (and, for that matter, most governments) routinely operate with substantial net borrowing. Of course, for every borrower, there must be a lender; organisations such as banks and insurance companies provide the balancing credit.

Here we consider 'liabilities proper', the current and long-term liabilities that, with shareholders' funds and minority interests, make up the total liabilities (see Equations 5.3, 5.4). There is some variation in the way in which different companies arrange this part of the balance sheet; for this and other reasons the classification of liabilities may be confusing. Fig. 5.2 shows how the various classes of liability are related, and introduces the more common names applied to them. It will be helpful to look back at this diagram from time to time while reading the descriptions that follow.

Long-term Liabilities

These fall into two broad categories: Creditors, amounts falling due after more than a year; and Provisions. As usual, we look in turn at the two kinds of contributions.

Creditors: amounts falling due after more than a year. These are the amounts owed by group companies. Two elements can be distinguished:

Borrowings
Other creditors

These relate to different aspects of the group's operations, as indicated in Fig. 5.2.

The amount entered as *Borrowings* represents the group's *Loan Capital*. This part of the capital employed comes not from shareholders, the nominal owners, but

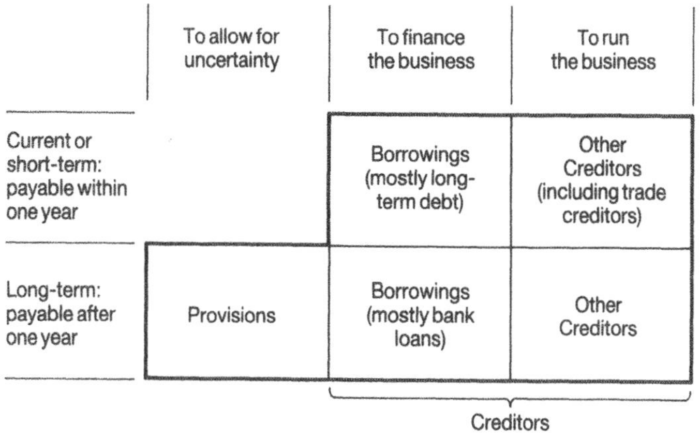

	To allow for uncertainty	To finance the business	To run the business
Current or short-term: payable within one year		Borrowings (mostly long-term debt)	Other Creditors (including trade creditors)
Long-term: payable after one year	Provisions	Borrowings (mostly bank loans)	Other Creditors

Creditors

Figure 5.2 Classification of liabilities

from outsiders who merely provide funds in return (they hope) for regular interest payments. Such debts may have terms of decades. *Finance Leases* come under this heading, since their effect (annual payments in return for the use of assets) is similar to that of interest-bearing loans used to purchase assets.

The *Other Creditors* heading relates more closely to the group's on-going business. Some elements are entered as 'long-term' liabilities simply because the bills happen to fall due more than a year into the future. Their current counterparts are essentially similar; they are analysed below (see Current Liabilities).

The *Loan Capital or Long-Term Debt* of a company is classified under headings such as:

Debenture loans
Bank loans and overdrafts
Amounts owed by group companies
Amounts owed by related companies

We need not be concerned with the exact distinctions between these forms of indebtedness, though they become crucially important when a company gets into financial trouble. Certain kinds of debt can be called in by the lender, and this can bring down the company through a shortage of cash. Some are secured against specific assets of the company, say one of its buildings; others are not.

The rates of interest on debts vary considerably, depending on the time at which the loans were taken out. Groups often arrange loans in the countries in which their subsidiaries are based or trade, in order to insulate themselves to a degree from the effects of changes in currency values.

The indebtedness of related companies, and specifically joint ventures, is not always shown clearly in group accounts. *Off-Balance-Sheet Financing* is a matter of some concern to the accounting profession. An example of such financing is provided by the announcement in the autumn of 1991 that Babcock International plc and Scottish Power plc were to set up a joint venture, Hampshire Waste to Energy Ltd. This company was to build, own and operate a 38 MW power station to be fuelled by Portsmouth's municipal waste. The £105 million required to fund this project would be borrowed by the joint-venture company and need not be recognised on the balance sheets of the partners.

Provisions. These are sometimes given the extended name *Provisions for liabilities and charges*. They are payments that are likely to be necessary; the uncertainty expressed by the word 'likely' relates to the amount that may have to be paid, to the time of payment, or to the probability of payment ever being necessary.

For companies operating in the United Kingdom the usual categories of provisions are those relating to pension benefits and to deferred taxation. The latter is discussed in Chapter 8.

In other countries, notably the United States, significant provisions are required to allow for potential health-care benefits for employees and pensioners. Other reasons for introducing provisions into the balance sheet are difficulty in remitting income from overseas subsidiaries, warranty obligations to purchasers of the group's products, and anticipated costs of discontinuing businesses.

Provisions appear on the liabilities side of the balance sheet, and are set off against unspecified elements among the assets. The balance sheet tells that at some time more than a year in the future Assets will probably have to be realised in order to discharge the Provisions.

By creating provisions, accountants have opportunities to tidy up the balance sheet at the end of the reporting year, and to prepare it for events anticipated for the years to come. The plausibility of guesses about future events decreases as one tries to look further ahead. Hence some categories of provision (notably deferred taxation) cover only expenditure expected within say three years. This approach is not applicable to pensions, of course, where actuarial tables provide a sound basis for predictions far into the future.

It is interesting to note that the most common provisions, those relating to pensions and deferred taxation, are equivalent to interest-free loans to the company. Rather than drawing the whole of their emoluments as they are earned, employees leave their pension benefits with the company. By accepting payment in arrears, tax authorities allow the company to use the tax due for considerable periods of time.

Current Liabilities

These may be labelled *Creditors: amounts falling due within one year*. The amounts owed by group companies go under a rather confusing array of names, some being:

Payments received on account
Trade creditors
Bills of exchange
Accruals and deferred income
Taxation and Social Security payments
Dividends due

In addition, there may be items under any of the classes of Long-term Liabilities, simply because their due dates do not come within the rather arbitrary one-year horizon. Conversely, some of the liabilities in the categories listed above may not fall due for more than a year, and they are labelled Long-term Liabilities. Thus the sorting of liabilities into post-one-year and pre-one-year categories has the effect of mixing up the borrowings associated with the long-term financing of the business, with the shorter-term indebtedness arising from its operations.

Most of the elements listed above are either obvious (for example, Trade Creditors are simply suppliers who have not yet been paid), or too complicated for us to worry about here. Two items do require comment. Both arise from the application of the Accruals Concept in drawing up the accounts.

Much of the *Corporation Tax* which a company ultimately pays on its year's profits is not demanded until well into the following financial year. At the balance-sheet date the company has a clearly defined obligation to pay this tax. Similarly, the *Dividends* that will be paid to shareholders are 'declared' or announced well in advance of the date on which the cheques are sent out. Here too there is a clear obligation to make payments during the next financial year.

Example 5.3 Trafalgar House PLC: 1990: Analysis of Liabilities

Example 5.2 dealt with the assets of this group; we look now at the other side of its balance sheet. Borrowings are shown in Table 5.7; debts of the holding company are also listed in a Note to the accounts, not reproduced here. Of the total shown, £152.2 million is said to fall due within one year. The word 'secured' implies that the lenders have a charge upon certain assets of subsidiary companies. Such security may cause them to lend at slightly lower rates of interest.

Nineteen distinct Debenture Loans are identified in the Note, and are given various names: debenture stock, loan stock, loan notes, US$ notes, and bonds. These names may not signify fundamental differences. The rates of interest payable on these loans are between 4⅞ per cent and 10⅞ per cent. The dates of repayment extend to the year 2014.

In addition to the Borrowings listed above, Trafalgar House reports Other Long-Term Creditors of £20.4 million.

Trafalgar House does not tell us much about the Provisions that it has made, but they appear to relate in the main to pensions and deferred taxation. Accountants must be allowed their little secrets. The total provided comes to £128.3 million after unspecified additions, 'reclassifications' and 'exchange adjustments'. During the year £36.2 million was utilised, that is,

used to discharge liabilities that materialised in that period.

We are now able to identify, in Table 5.8, the several kinds of long-term liabilities of this group.

Turning to Current Liabilities, we find the values given in Table 5.9. The short-term Borrowings are mostly bank loans and overdrafts. The major elements of Other Creditors are payments in advance (£111.0 million), trade creditors (316.4), and accruals and deferred income (296.7). Corporation tax owed is £17.0 million, and other tax and social security payments (including some collected from employees) amount to £56.5 million.

Ordinary dividends of £48.3 million are noted among the Other Creditors. The Note on dividends reveals that the interim payment was £44.2 million, while the final dividend will be £48.3 million. As is customary, the final dividend had not been sent to the ordinary shareholders by the balance-sheet date.

Table 5.7   Classification of Long-Term Debt

Debenture Loans	
Secured	4.4
Unsecured	427.0
Bank Loans	
Secured	23.9
Unsecured	113.4
Bank Overdrafts	
Unsecured	56.6
	£625.3 million

Table 5.8   Calculation of Total Long-Term Liabilities

Borrowings	£625.3 million
Borrowing due within a year	(152.2)
Other Creditors	20.4
Provisions	128.3
	£621.8 million

Table 5.9   Current Liabilities

Borrowings	£152.2 million
Other Creditors	890.5
	£1042.7 million

PLATE 6

Dartford Crossing of the Thames, construction and operation by Trafalgar House.

5.4 Working Capital

We have already met this quantity, but passed it by without a formal introduction. In Equations (5.5) and (5.6) appear the Net Current Assets, the current assets less the current liabilities; 'Working Capital' is just another name for this difference. And a strange name it is, since this element comes on the assets side of the balance sheet, while the Share Capital (part of SHF) and the Loan Capital (part of long-term liabilities) are on the other side. Yet other names for this quantity are *Net Working Capital* and *Liquid Assets*.

Note that Debtors (a contribution to Current Assets) includes all payments expected both within and beyond the coming year. However, Current Liabilities includes only those Creditors due within the year. This biases the Net Current Assets – or Working Capital – upwards, but not usually to a significant extent.

Whatever name it is given, working capital is the part of the assets that turns over to keep the business running. Equation (5.5) can be expressed in words as:

Fixed Assets + Working Capital = Capital Employed

This statement expresses an important feature of any business enterprise, namely, that the available capital must be deployed in two ways:

a) in acquiring facilities with which and within which the business can be carried out; expenditure to this end is termed *Capital Expenditure*
b) in funding myriad day-to-day costs, including those of employees, services such as electricity, the bills of suppliers of materials and components, and payments associated with borrowing; this is the Working Capital.

A company can be asset-rich, but still fail through an inability to use those assets to generate cash rapidly enough to fund the numerous minor transactions that keep it running. In a period of rapid growth, more cash is drawn into the *Asset-Conversion Cycle*. In Fig. 1.3 this is the inner loop between the Production Unit and the Market.

Although balance-sheet quantities do not relate directly to cash passing into and out of the business, working capital gives us some indication of a company's present cash position and its likely position in a few months' time. Thus the possession of large Net Current Assets, relative to the total capital employed or to the annual turnover, is a sign of financial stability, suggesting that the company has the potential to pay its debts as they come due, even if trade turns sour.

On the other hand, capital that has been used to make goods for which payment has not been received can hardly be said to be 'working' effectively. It can be argued that even cash accruing interest in the bank is not earning as effectively as the shareholders might expect. Hence there is a balance to be struck: too high a level of working capital suggests inefficiency; too low a level may suggest potential instability. The appropriate level depends very much on the nature of the business, as is illustrated by the following example.

Example 5.4 British Aerospace plc 1990; Glaxo Holdings plc 1990;
J Sainsbury plc 1991: Net Current Assets

The reporting years of the three companies differ, but the periods considered here overlap. Two of these companies have been introduced in earlier examples. The third is Britain's largest and apparently most efficient retailer, with 12.6 per cent of the national trade in food and drink. It is included to give a broader perspective on company finance.

Table 5.10 lists the Current Assets and Liabilities of these companies, reduced to a common format. The Trade Debtors (from whom payments are expected) and the Trade Creditors (suppliers who expect to be paid) are identified explicitly in this table.

As might be expected in view of the very different activities of these companies, there are marked differences in the composition of Net Current Assets. Although British Aerospace and Glaxo Holdings report nearly identical values for NCA, the quantities that are subtracted are very different. BAe has large stocks, including much work in progress, and owes a great deal to its trade creditors. Glaxo's stocks are modest and its debts to trade creditors almost negligible.

Sainsbury's stock level is modest, but it owes to trade creditors much more than is owed to it. Accordingly, this company has a negative value of Net Current Assets and hence negative Working Capital. Neverthe-

Table 5.10 Contributors to Net Current Assets (in £ million)

	BAe	Glaxo	Sainsbury
Current Assets			
Stocks	2830	392	361
Trade Debtors	900	463	19
Other Debtors	420	313	126
Investments	1493	1552	–
Cash	366	20	111
Total Assets	6009	2740	617
Current Liabilities			
Trade Creditors	1725	123	556
Prepayments	1364	–	--
Other Creditors	1536	1212	873
Total Liabilities	4625	1335	1429
Net Current Assets	1384	1405	(813)

less, in view of its sound long-term trading performance and enormous cash through-put, Sainsbury must be regarded as a very stable company.

In Table 5.11 we look at current assets in other ways. Here are presented the Liquid Funds (the sum of cash and short-term investments), the NCA less these Liquid Funds, and the difference between Trade Debtors and Creditors. Unlike the other two, Sainsbury maintains a very low level of liquid funds, apparently seeing opportunities to invest in its business all of the cash that comes to it.

By deducting Liquid Funds from Net Current Assets, we obtain a better representation of the funds actually absorbed in each business. Viewed in this way, these three large companies all appear to be operating in debt, that is, owing rather more than they are owed.

Yet another view of the finances of these companies is provided when we look at the Net Trade Debt, that is, the difference between Trade Debtors and Trade Creditors. Apparently Glaxo pays its bills quickly, while the other two manage to be paid themselves before paying off their own short-term debts. It is not difficult to see how this comes about, particularly in the case of the grocery company, Sainsbury.

These comparisons reveal some of the special features of the finances of an engineering company such as British Aerospace, many of whose products can be classified as capital goods. Unlike the other companies, it has received substantial Prepayments from customers for work it is currently undertaking. Indeed, prepayments account for most of the Liquid Funds of the group.

Table 5.11 Other Aspects of Current Assets (in £ million)

	BAe	Glaxo	Sainsbury
Liquid Funds	1859	1572	111
NCA less Liquid Funds	(475)	(167)	(924)
Trade Debtors			
less Trade Creditors	(825)	340	(537)

5.5 Shareholders' Funds

A shorter equivalent term is *Equity*, but the more cumbersome Shareholders' Funds is commonly used in Britain, unless shortened to SHF. The funds invested by shareholders, in various ways and at various times, are listed in the balance sheet under such headings as:

Share capital
Share premium account
Revaluation reserve
Other reserves
Profit and loss account

The last four are collectively called Reserves.

Share Capital

To understand what is meant by *Share Capital* and *Share Premium Account*, it is necessary to know that shares are assigned a *Par Value*, usually a small value unrelated to the current market value or indeed to the *Issue Price* at which the company originally sold the shares. For example, in 1987 Her Majesty's Government sold its ordinary shares in Rolls-Royce plc for 170 p each, although the Par Value assigned to the shares was 20 p. Each share issued contributed £0.20 to the Share Capital, while adding £1.50 (that is, 170 p less 20 p) to the Share Premium Account. The shares of some overseas companies do not have a Par Value, but those companies seem not to be handicapped by this deficiency.

Strictly speaking, it is only *Called-Up Shares* that contribute to the SHF. These are shares that have been paid for or for which payment has been received or demanded. This component of the balance sheet is sometimes labelled more precisely as *Called-up Share Capital*.

Directors can, having been given authority by a general meeting of shareholders, make a *Capitalisation Issue* of ordinary shares to the existing shareholders. This may also be called a *Scrip Issue* or a *Stock Dividend*, and the shares issued may be called *Bonus Shares*. No payment is received for these shares, and the procedure merely transfers funds from the Share Premium heading to the Share Capital heading. This does not make the company any more valuable (indeed, it is poorer by the costs of printing and posting share certificates), but it has a number of technical advantages.

In Chapter 6 we shall consider other features of shares, and things that are done with them.

Reserves

It is only in a technical sense that the *Share Premium Account* is distinct from the Share Capital: it is their sum that represents the funds actually subscribed for the company's shares when they were issued to their first owners. In Section 3.2 this sum was referred to as Share Investment.

The *Revaluation Reserve* is easy to understand. When a company revalues some of its assets, in order to make its balance sheet represent their current values more realistically, a balance must still be maintained. This reserve is created on the liabilities side of the balance sheet to compensate for the (presumed) increase in assets on revaluation.

The *Profit and Loss Account* is the repository for the annual transfers calculated in the series of annual profit and loss statements. But that is not the only way of changing this particular reserve (for it is a reserve despite its different name). In Section 5.6 we shall see that this Account, or others of the reserves, can be affected when subsidiaries are bought and sold.

The creation of *Other Reserves* is very much at the discretion of boards of directors and their advisors. Some names that may be encountered are:

Reserves of subsidiary companies
Reserves of associated companies
Capital reserve
Special reserve

The nomenclature adopted is not important, but it is worth noting that the Articles of Association of a company may define certain reserves as *Undistributable*. This means that the directors are not free to dip into them to pay a dividend. For a company in good financial health this distinction is not important. But not all companies have that happy characteristic.

The possession of large 'reserves' does not mean that a company has lots of cash on hand or can readily put itself in that position. All that we know is that the company has earned much money in the past; whether it still has it is another matter.

The term Shareholders' Funds embraces reserves created by retentions from annual profits as well as funds actually paid in by shareholders. In fact, retained profits are by a considerable margin the greater part of the Equity of most well established companies. The lumping of 'active' and 'passive' inputs of capital implies that the shareholders are willing participants in the decision to retain a considerable part of 'their' Earnings for the use of the company. The justification for this assumption is not entirely clear.

Example 5.5 Trafalgar House plc 1990: Classification of Reserves

---

We have looked at most other aspects of the group balance sheet of this company, and the picture is rounded off by the list of reserves in Table 5.12. In Table 5.3 we identified the source of the Revaluation Reserve, namely up-dated values for the group's hotels and business premises. From the Note giving details of changes in

Table 5.12 Calculation of Total Reserves

Revaluation reserve	£ 84.0 million
Special reserve	115.1
Additional reserves	176.0
	£375.1 million

reserves over the year it can be deduced that the Special Reserve responds to the buying and selling of subsidiaries. The Other Reserves heading includes the Profit and Loss Account; unusually, it is not specifically identified in this balance sheet.

In Table 5.13 the Shareholders' Funds are determined. The total can be reconciled with the Net Assets of £840.9, as shown in Table 5.14. Adding Minority Interests of £14.7 million, we find the Capital Employed to be £840.9 million.

Table 5.13   Calculation of Shareholders' Funds

Called up share capital	£127.1 million
Share premium account	324.0
Total reserves	375.1
	£826.2 million

Table 5.14   Calculation of Net Assets

Fixed Assets (Example 5.2)	£723.4 million
Net Current Assets (Examples 5.2, 5.3)	739.3
Long-term Liabilities (Example 5.3)	(621.8)
	£840.9
Minority Interests	(14.7)
	£826.2 million

PLATE 7

An offshore drilling module for BP, completed by Redpath Offshore, a Trafalgar House subsidiary.

5.6 Changing the Balance Sheet

General Principles

The balance sheet is a snap-shot of the company's financial position at one instant; we have so far looked at that instantaneous condition without giving much thought to how it arose. The fundamental idea to be kept in mind when thinking about changes in the balance sheet is that it must always remain in balance.

We noted earlier that the revaluation of assets cannot be divorced from an adjustment of a revaluation reserve to keep the total assets and liabilities in step. Moreover, the incorporation of a year's (positive) retained profit into the profit and loss account must be balanced by increases in some of the assets of the group, or possibly by reductions in some of its liabilities. Indeed, each line of the profit and loss statement implies a set of compensating changes on the two sides of the balance sheet.

Consider for example the line in the PowerGen's profit and loss statement (see Example 4.2) that reads:

Net interest receivable   £59.8 million

This implies entries on the balance sheet under the headings:

- Cash (an asset): payments both received and paid out during the year covered by the report
- Debtors (an asset): payments not yet received but expected in the coming year
- Creditors (a liability): payments to be made in the coming year

Moreover, by the reporting date the net cash received will have been disbursed in a variety of ways. Thus the Net Interest affects numerous ledgers that track both assets and liabilities.

When a depreciating asset is sold, let us say for a price above its book value, that book value vanishes from the fixed assets, while the liquid assets are increased by the sale price. To keep the liabilities in step, an exceptional or extraordinary item is introduced to pass the gain through to the reserves. The asset will continue to be depreciated on the balance sheet of the acquiring company, but the depreciation will be based on the price at which it was purchased by that company. In the case envisaged, the charge may well be higher than before the transfer of ownership took place.

Accounting for Goodwill

Every decade or so, a mild frenzy grips the corporate community, with many holding companies seeking to expand their empires by acquiring new subsidiaries. Seemingly, boards of directors accept in part the words with which Jane Austen begins *Pride and Prejudice*:

> "It is a truth universally acknowledged, that a single man in possession of a good fortune, must be in want of a wife".

Many corporate marriages arranged in haste are repented of, not in leisure, but in frantic efforts to avoid the consequences of a balance sheet overloaded with debt. Miss Austen's point that it is necessary to be "in possession of a good fortune" is not always understood by boards of directors.

The possibility of claiming ownership of an intangible asset called Goodwill was introduced in Section 5.2, where it was defined as the difference between the purchase price of a subsidiary and the value of its identifiable assets. We shall now see how this short-fall can be entered into the balance sheet; that is to say, how it is possible to account for the missing assets. Two very different procedures are considered below.

Writing Off Goodwill. Here the premium of purchase price over assets is simply subtracted from reserves, usually from the profit and loss account. This has the required effect of reducing the total liabilities of the group to match the fall in its total assets. These operations have the apparently anomalous effect of reducing the shareholders' equity, even though they now own a larger group.

Writing-off has been the standard procedure in the United Kingdom, although there is reason to believe that the method described below will supersede it before long. The procedure now used has little effect on the profit and loss statement for the year in which the purchase is made. Effects do develop there in later years, possibly through an increase in interest outgoings (if the purchase requires added debt) or through a decrease in earnings per share (if more shares are issued to pay for the purchase). Of course, if the purchase is judicious, the additional profit it generates will carry the interest payments and sustain the earnings per share.

Creating an Intangible Asset. Here assets equal to the whole of the purchase price are introduced somewhere among the assets. In particular, the Goodwill appears as an intangible asset.

This method affects the profit and loss statements for subsequent years in an additional way, namely through the notional cost of *Amortisation* of the intangible asset. Amortisation is the special name given to depreciation of Goodwill; typically, this cost is spread over a period of from twenty-five to forty years. This additional cost makes the financial consequences of the purchase more obvious, and this may act as a disincentive to acquisition.

This approach has been adopted for many years in the United States and seems likely to become standard practice in the United Kingdom as well. In Example 3.2 we noted that the Ford Motor Company had created an intangible asset recognising the Goodwill element in its purchase of Jaguar plc and other subsidiaries. In earlier years Ford UK had written off Goodwill from reserves; perhaps the change of policy was adopted to simplify the consolidation of its accounts into those of its American parent.

A Mixed Approach. Some companies argue that specific aspects of newly acquired subsidiaries warrant recognition as intangible fixed assets. They need then write off only part of the Goodwill on purchase.

A number of food companies have argued in recent years that well-established brand names acquired with purchased subsidiaries can quite properly be introduced on the balance sheet as intangible assets. There is some merit in this argument, for the very fact that a price has just been agreed for the subsidiary company allows the calculation of a specific value for its intangible assets. In the same way Ford could argue that the Jaguar marque has a market-determined value.

The accounting profession is divided on a related issue, whether 'home-grown' brands can also be assigned values and be recognised on the balance sheet. Yet another contentious issue is whether the value of a brand should be amortised, or whether it should retain its initial value (or even be augmented) in recognition of subsequent advertising promotion and product improvement.

Merger Accounting

The techniques just considered can be classified as forms of the *Acquisition or Purchase Method* of fitting together companies' accounts. A less commonly adopted approach is the *Merger or Pooling Method*. Here the entire year's accounts of the merging companies are pooled, no matter at what point in the year the merger took place. The accounts for earlier years are also pooled to provide a basis of comparison. Payment normally takes place by exchange of shares, and there is no need to identify the purchasing and the purchased company. Goodwill is not recorded on the pooled balance sheet, and the question of amortisation does not arise. These procedures are generally believed to present a less realistic picture of the effects of merger than that provided by the more widely adopted Acquisition Method.

Example 5.6 Effects of Various Ways of Paying for an Acquired Subsidiary

---

To illustrate ways of fitting the assets and liabilities of a new subsidiary into the balance sheet of the acquiring group, we consider a hypothetical example. A group acquires a subsidiary for a purchase price of £250 million. Before the acquisition takes place, the balance sheets of the group and acquired company are as shown in Table 5.15.

Table 5.15 Balance Sheets of the Two Companies

	Initial Balance Sheet of the Group		
Tangible Assets	1250	Share Capital	500
Net Current Less Cash	250	Reserves	500
Cash	250	Long-term Debt	750
Net Assets	1750	Capital Employed	1750
	Balance Sheet of the Acquired Company		
Fixed Assets	100	Share Capital	50
Net Current Less Cash	25	Reserves	–
Cash	25	Long-term Debt	100
Net Assets	150	Capital Employed	150

We note that:
a) The group Shareholders' Funds are £1000 million before the acquisition.
b) The Shareholders' Funds of the acquired company are £50 million before acquisition.
c) The Goodwill involved in the purchase can be calculated as:

Purchase Price − Identifiable Value
= 250 − (150 − 100) = £200 million

We shall use Acquisition Accounting to determine the form of the group balance sheet after acquisition, considering five possible ways of funding the purchase:

Case A. Cash; Goodwill written off

Case B. Cash; Intangible asset created
Case C. Issue of Debentures; Goodwill written off
Case D. Issue of Shares; Goodwill written off
Case E. Sale and Lease-Back of Fixed Assets; Goodwill written off

The balance sheets resulting from these procedures are set out in Tables 5.16 to 5.20. It will be instructive to trace through the various balance-sheet adjustments. Note that the post-acquisition Capital Employed and Shareholders' Funds depend on the purchase method that is adopted.

Table 5.16   Balance Sheet following Purchase for Cash (Goodwill written off)

Tangible Assets	1350	Share Capital	500
Net Current Less Cash	275	Reserves	300
Cash	25	Long-term Debt	850
Net Assets	1650	Capital Employed	1650

Table 5.17   Balance Sheet following Purchase for Cash (Goodwill recognised as Intangible Asset)

Intangible Assets	200		
Tangible Assets	1350	Share Capital	500
Net Current Less Cash	275	Reserves	500
Cash	25	Long-term Debt	850
Net Assets	1850	Capital Employed	1850

Table 5.18   Balance Sheet following Purchase funded by Issue of Debentures (Goodwill written off)

Tangible Assets	1350	Share Capital	500
Net Current Less Cash	275	Reserves	300
Cash	275	Long-term Debt	1100
Net Assets	1900	Capital Employed	1900

Table 5.19   Balance Sheet following Purchase funded by Issue of Shares by Parent Company (Goodwill written off)

Tangible Assets	1350	Share Capital	750
Net Current Less Cash	275	Reserves	300
Cash	275	Long-term Debt	850
Net Assets	1900	Capital Employed	1900

Table 5.20   Balance Sheet following Purchase funded by Sale and Lease-Back of Assets (assuming sale at a premium of 25 per cent to book value; Goodwill written off)

Tangible Assets	1150	Share Capital	500
Net Current Less Cash	275	Reserves	350
Cash	275	Long-term Debt	850
Net Assets	1700	Capital Employed	1700

5.7 More Words of Warning

We earlier noted some of the problems associated with inflation (Section 4.6) and with the interpretation of a group's profit and loss statement (Section 4.7), as well as more general difficulties in applying the basic accounting principles. This does not complete the list of difficulties in balance-sheet interpretation.

The choice of the *Balance-Sheet Date* can be significant. For a company whose business has a marked annual cycle, the accounting period should be exactly one calendar year, if a realistic view is to be obtained. Consider for example a distributor of greeting cards: a shift of reporting date by a few weeks relative to the Christmas rush could change the balance sheet values profoundly. The automobile business, in the United Kingdom in particular, is also subject to swings in production and in balance-sheet structure, as the traditional August rush approaches and passes.

The longer-term and less predictable business cycles affect financial statements profoundly as the level of activity waxes and wanes. These effects are also strongly dependent on the balance-sheet date.

A company is able to select for the end of its accounting period a date on which its balance sheet is usually in good shape, that is a point at which it can expect to have received payment for much of the previous year's output. The accounts will then show low current debtors and either reduced borrowings or a high level of cash. It is not impossible for a company to 'massage' its balance sheet as the reporting date approaches. This might be done through a crash programme of invoicing goods to reduce stocks and increase reported turnover. Or it might be done by a short-term financial transaction across the year end, temporarily reducing borrowings.

We noted earlier that borrowing by joint ventures and other associated companies may not be fully recognised in group accounts. Note also that a company is not required to consolidate the accounts of subsidiaries whose business is of a distinctly different character from that of the group as a whole. In the past this has allowed holding companies to avoid reporting the affairs of financial subsidiaries, in particular. This made it possible for the group balance sheet to be 'improved' by the sale of its accounts receivable to a captive finance company, in return for cash. (The borrowings of the financial subsidiary did not have to be shown in the group accounts.) The Creditors item in the group accounts is thus reduced, and Cash appears on the balance sheet in its place. Such practices are being curtailed by increasingly stringent reporting standards.

The nature and extent of 'hedging' and other currency management practices is not usually made explicit in group accounts. This is something of a relief to most users of the accounts, for these matters are doubtless difficult to explain and to comprehend. But it is also an area of substantial doubt.

Significant acquisitions or divestments give rise to substantial changes in both the assets and the liabilities of a group. Often they are accompanied by a programme of revaluation of assets joining or leaving the group. There are plentiful opportunities for earlier mistakes – either of accounting or business judgement – to be submerged in the flux. Moreover, it may be difficult to determine the profits and cash generated by trading in the year of transition.

Writing in the *Financial Times* in September 1991, John Plender commented as follows on the effects of acquisitions:

"The extreme latitude offered by British acquisition accounting and flabby auditing means that any UK predator can legitimately generate unreal profits by recording the target company's assets in its books at so-called fair (ie: unfair and artificially depressed) values. Subsequent sales, whether of goods or subsidiaries, can then yield artificially high profits."

We must hope that more rigorous standards of financial reporting will curb these practices.

6

Dealing with Shares and Stocks

In Chapter 5 the concepts of Share Capital and Loan Capital were introduced, but little was said about how they come into being or about their roles in the life of a company. Here we consider these matters, primarily from the point of view of the company, but adopting also the position of the investor, whose funds the company must attract to develop and sustain its activities. We look first at ordinary shares, the source of most share capital, and then at other securities issued by companies.

6.1 The Characteristics of Shares

Ordinary Shares

It may seem anomalous that the most important class of shares are called *Ordinary Shares* and in other countries *Common Shares* or *Common Stock*. As the features that distinguish them from other securities are explained, it will become apparent that such shares offer to the investor a complex of benefits and risks. It is the interplay between these attributes that generates rapid, often large changes in the market value of ordinary shares.

Dividends. While ordinary shareholders cannot be completely confident of receiving dividends, in practice most established companies provide a stream of payments that seldom decrease from one year to the next. In a bad year the directors of a company may choose to raid its distributable reserves, rather than reduce dividends or pass them altogether. An *Interim Dividend* is usually paid during the year and a *Final Dividend* (usually larger) during the next accounting year, following approval by shareholders at the Annual General Meeting.

Liquidity. For most companies ordinary shares represent much the greatest part of the share capital. For this and other reasons they are the most actively traded of company securities and thus offer the greatest liquidity. However, the price that can be obtained is somewhat uncertain. It fluctuates from day to day, sometimes violently, and occasionally trading in a company's shares is suspended by a stock exchange.

Rights of Owners. Those registered as owners of ordinary shares are the theoretical owners of the company that issued them. The shareholders have a measure of control over the affairs of the company through their voting rights (proportional to the number of shares held) on all issues crucial to its future and on many routine matters as well. In practice, the control of a company with widely dispersed shareholdings does not lie with the mass of shareholders, but with the directors or professional managers employed by the company, or perhaps with a small number of large shareholders.

Some companies issue *Non-voting Ordinary Shares* in addition to their voting counterparts. The former usually trade at slightly lower prices, since the control element is absent from their value. The issue of such shares is not encouraged by the Stock Exchange.

If a company is liquidated, the owners of its ordinary shares receive all the proceeds left after the winding up, once the true liabilities have been fully discharged. Ordinary shareholders are also last in the queue when the annual profits of the company are distributed. Thus they have no guarantee of repayment, either annually through dividends or through the return of the capital invested.

Issuing Shares. At any given time the directors of a company will have the authorisation of the shareholders to issue a specified number of shares. This number, multiplied by the par value of the shares, defines the *Authorised Share Capital*. Following an offer for sale, the directors will allocate shares to those who sought to purchase them; these are *Allotted* shares. They are subsequently *Issued* to shareholders and later, when payment has been demanded, become *Called-Up* shares. If payment is by instalments, the shares are for a time *Partly Paid*, before reaching the final status of *Paid-Up* or *Fully Paid* shares. At each stage the share capital may change. What is reported on the balance sheet is the *Called-Up Share Capital*.

Preference Shares

In North America these shares are called *Preferred Stock*. They have 'preference' by ranking ahead of ordinary shares in their right to receive dividends, usually at a level fixed throughout the life of the shares. What is more, if the company is wound up, this class of shareholders receive a share of the realisable assets before the ordinary shareholders, but after holders of debt. Payment is limited to the nominal value of the preference shares.

Some classes of preference shares are labelled *Cumulative*, indicating that missed dividend payments are rolled up, for payment when the company's profits recover. Another adjective applied to preference shares is *Redeemable*. This indicates that the shares may be bought back at certain times, at the option of either the company or the shareholder. All sorts of conditions can be attached to the option, for example, the requirement that it must be exercised on a certain date within a band of years.

The interest carried by preference shares reflects the rates typical at the time at which they were issued, while the market price of any particular series of shares is dependent on the fixed interest it pays. Shares issued many years ago – when interest rates were around 5 per cent, say – will now commonly trade at about one-half their nominal value, since fixed-interest securities have in recent years offered around 10 per cent interest. There is little attraction to a company in buying back such low-interest shares, since the cost of the funds they represent is significantly lower than that of new borrowings. However, if such shares are redeemable at the option of the holder, the market price will rise as the redemption date approaches, since the benefits of ownership include the redemption value as well as the interest that will be paid prior to redemption.

Convertible Shares or Stock

Like preference stock, this is a hybrid security, with some of the characteristics of ordinary shares and some features of long-term debt. It starts life as a loan, with fixed interest, but may end up as ordinary shares, through the exercise of the holder's option of *Conversion*. Again a variety of conditions may attach to the option, the most important being the price to be paid for ordinary shares on conversion.

When such stock is converted, the total number of ordinary shares in issue increases, and the fraction of the earnings attributable to each share falls. Hence companies often present two values for the Earnings per Share ratio that was introduced in Section 4.3. One is based on the number of shares currently in issue. The other, called the *Diluted or Fully Diluted EPS*, is based on the number of shares that would exist if all the existing options to convert were exercised.

Options and Warrants

In the remuneration packages of directors and employees, a company may include the right to purchase shares at specified prices, but at some time in the future. The Fully Diluted EPS allows for the assumed exercise of all of these *Options*. They are not transferable, being of use only to those to whom they are granted. Their value to the recipient depends on the market price of ordinary shares at the time at which the options can be exercised. Hence they provide an incentive to improve the performance of the company, and thus to increase the price of its shares.

Warrants are more generally exercisable options to subscribe for ordinary shares at a specified price. They are traded, since ownership can be transferred. They too dilute the existing shareholdings when exercised. The market price of warrants is highly volatile, since it is linked to the difference between the present market price and the exercise price.

It will now be apparent that the share capital of a company is a complex entity, comprising not only ordinary shares, but instruments that possess some characteristics of loans, and with the potential for future change implicit in options granted to those associated with the company and to outsiders who have purchased warrants.

Example 6.1 Westland plc 1985 to 1990: Forms of Share Capital

---

This is Britain's major manufacturer of helicopters. It also makes structural components for aircraft, specialising in composite materials, and a variety of environmental-control systems for aircraft. In the past Westland was the mainstay of the British hovercraft industry, though that business has fallen away in recent years.

On the 30 September 1985 Westland's authorised share capital was £20 million, divided into 80 million ordinary shares with a par value of 25 p each. Only 59.3 million shares had been allotted and issued, giving a called-up capital of £14.8 million.

On the 30 September 1990 the share capital was very different, so different that a table is used to display its features; the number of shares in each class is shown in brackets. In addition, there were 24.2 million warrants outstanding, each carrying the right to subscribe for one ordinary share at 81.7 p per share on the 31 January in each year up to and including 1996.

Under the SAYE (Save-As-You-Earn) Share Option Scheme, 2.45 million options

had been granted to full-time employees. The option prices varied between 72 p and 140 p. This tax-saving scheme provides for purchase either five years or seven years after the options are offered. Under the Directors' and Senior Executives' Share Option Schemes, 2.81 million options had been granted, at prices between 80 p and 155 p. The periods during which these options can be exercised extend to the year 2000.

The calculation of the Earnings per Share is explained as follows:

"The basic earnings per Ordinary Share is calculated on profit attributable to ordinary shareholders of £15.7 million after deducting a preference dividend of £3.5 million and on a weighted average number of 126.0 million allotted Ordinary Shares.

"Fully diluted earnings per Ordinary Share is calculated on a notional profit of £18.7 million and allotted Ordinary Shares totalling 180.2 million to allow for the full exercise of all options and conversion rights."

The kinds of restriction imposed on con-

Table 6.1 Called-up Share Capital 1990 (£ million)

	Authorised Capital	Allotted and Issued
Ordinary Shares of 2.5 p (533.8 million authorised, 91.1 allotted and issued)	13.3	2.3
Voting Preferred Ordinary Shares of 2.5 p (500,000)	0.00	0.00
Non-voting Preferred Ordinary Shares of 2.5 p (35 million)	0.9	0.9
7–1/2% Convertible Cumulative Preference Shares of £1 (20.262 authorised, 20.258 allotted and issued)	20.3	20.2
10% Cumulative Redeemable 'A' Preference Shares of £1 (2 million)	2.0	2.0
10% Cumulative Redeemable 'B' Preference Shares of £1 (18 million)	18.0	18.0
Special Deferred Ordinary Shares of 2.5 p (30.6 million)	0.8	0.8
	55.3	44.2

version are illustrated by the specifications given in a Note on Westland's share capital:

"The Non-voting and Voting Preferred Ordinary Shares can be converted into Ordinary Shares, at the holders' option, at any time prior to 31 January 2006. The Non-voting Preferred Ordinary Shares can be converted, at the holders' option, into Voting Preferred Ordinary Shares at any time.

"The 7½% Convertible Cumulative Preference Shares can be converted, at the holders' option, on 31 January 1996 in each year up to and including 31 January 2006 on the basis of 20.8 Ordinary Shares for every £17 nominal of Convertible Cumulative Preference Shares.

"The 10% Cumulative Redeemable 'A' Preference Shares are to be redeemed between 31 January 1994 and 31 January 1996 inclusive and the 10% Cumulative 'B' Preference Shares between 31 January 1994 and 31 January 1998 inclusive."

It would not be fruitful to go through all the stages by which the 1990 Share Capital was attained. It will be evident that these conditions are intended to achieve a balance between the desire of the providers of new capital for a measure of confidence that they will be rewarded, and the need for the company to be confident that it will not experience unpredictable demands for the repayment of the new capital. The original owners, the holders of 1985 ordinary shares,

had a much reduced stake in the company after this 'financial reconstruction'.

It is natural to wonder why this dramatic alteration to the share structure occurred. The surrounding events were indeed dramatic, involving the resignation of both the Defence Secretary and the Home Secretary from Her Majesty's government. An extract from the 1986 Chairman's Statement gives a sanitised version of these events:

"The first half of the financial year under review was dominated by the unfortunately delayed financial reconstruction of your Company. . . . This resulted in an injection of £75 million of new capital and the establishment of our formal association with United Technologies Corporation (UTC) and Fiat, the two strong international partners that we had considered essential to the long-term stability of the Company. I will not attempt to recount the problems encountered during the period leading up to this reconstruction: they have already received far too much publicity, much of which was thoroughly unhelpful to the company's international standing. At the end of the long-drawn-out struggle a private sector solution to our problems was achieved at no exceptional cost to the taxpayer. Your Company is now adequately capitalised, soundly

PLATE 8

E101 Heliliner, developed jointly by Westland and Agusta of Italy.

based and its future more secure. The management of the Group has been reorganised into four divisions. An initial and essential cost reduction programme was carried out and new product and marketing strategies have been defined."

In 1991 Westland is still independent, but its share register has changed markedly from the pattern established by the refinancing exercise of 1986. By 1991 GKN plc owned some 21.2 per cent of the ordinary shares and nearly half the Voting Preferred Ordinary Shares. Another significant shareholder was United Technologies, and Westland reported that GKN and UT "have given notice that they are to be regarded as acting in concert in respect of their holdings in the Company".

6.2 *Valuing a Company or Group*

There are a number of reasons for wanting to know the value of a company. The owners may wish to sell it. Someone may be interested in buying it. An investor needs to decide whether to hold the company's shares or loan securities.

Market Capitalisation

If the company's shares are traded actively, one measure of value is immediately available, the *Market Capitalisation*. This is just the current market price per share multiplied by the number of shares in issue. In a sense this is the 'real' value, for it can be argued that something is worth only what a buyer is willing to give for it. However, several other measures of a company's value to its present and potential owners can be distinguished:

- the Equity or Shareholders' Funds stated in the last accounts, often referred to as the Book Value of the company
- the *Liquidation Value*, which might be achieved by selling the assets individually and then discharging the liabilities
- the *Break-Up Value*, an estimate of the sum that might be achieved through *Demerger*, that is, the progressive sale of subsidiaries as going concerns, carrying their indebtedness with them, and
- a *Bid Value*, the amount that has been or might be offered for the shares of a company by a person or organisation that wished to gain control of it.

The share price and capitalisation depend primarily on expectations of future earnings, dividends and share prices. However, in some degree the capitalisation reflects the alternative measures of a company's value, through the perceived probability of its being liquidated, broken up, or bid for.

It can be assumed that those contemplating making a bid will consider all of these valuations, before deciding on their bid price. Equally, those receiving an offer of purchase will need to look at their investment in every possible way, before deciding how to react. Both parties will wish to consider intermediate solutions, involving partial demerger or partial liquidation.

It should be noted that other possible measures of a company's value, its Total Assets and Net Assets (or Capital Employed), are partly attributable to short-term and long-term creditors, who are not 'members' of the company. Hence there is little logic in using those quantities as estimates of its value.

Alternative Valuations

Shareholders' Funds. It has become clear that the equity component of the balance sheet is usually not a realistic estimate of the value implicit in the ownership of shares, as a consequence of writing-off goodwill and failing to revalue assets regularly and systematically. The inaccuracy of asset valuation is particularly telling, since SHF is the difference between the total assets and the total of the liabilities proper. Some elements of the latter, especially provisions, are themselves subject to uncertainty.

A few examples will demonstrate the weakness of the link between market and book values. During 1990 the market price of the ordinary shares of British Aerospace varied between 474 p and 607 p. As the average number of shares in issue was 257.7 million, the market capitalisation varied during the year between £1220 and £1560 million. The shareholders' funds were about £2450 million on the balance-sheet dates at each end of the year.

During 1990 the price of the shares of the BOC Group varied between 440 p and 611 p, and the number of shares in issue was around 467.6 million. Thus the capitalisation varied from £2060 to £2860 million. The SHF on the balance-sheet date that fell within this calendar year (30 September 1990) was £1774 million.

On comparing the values given in the last two paragraphs, one might wonder if some have been transposed. The SHF of British Aerospace happens to fall within the capitalisation range of the BOC Group, while the SHF of BOC is little more than the capitalisation of British Aerospace. In one case the SHF is about twice the market value; in the other the market value is about twice the equity.

Liquidation and Break-Up Values. These arise from two quite distinct models of the dissolution of a group. Liquidation is seen here as a more-or-less forced sale of assets, not necessarily embedded within operating companies. From the balance sheet one can obtain an estimate of the *Liquidation Value* or, in reality, a number of estimates for different assumptions about the market values of the group's assets. In the absence of a revision of the asset values, the balance sheet will simply suggest the SHF or equity as the amount achievable on liquidation. This estimate could be much too low: the book values of assets might have fallen far behind changing market values. Or it could be much too high: the accumulated depreciation might not be adequate to compensate for a forced sale of assets, subsequently no longer to fulfil their intended purpose.

The *Break-Up Value* of a group is estimated by imagining an orderly demerger of subsidiaries and sale of holdings in related companies. In principle, this could be done by issuing shares in each of the subsidiaries to existing holders of parent-company shares. When trading in the shares of the separate businesses begins, each has a distinct market capitalisation, and their sum is the Break-Up Value.

BAT Industries plc went part-way down this route, by demerging a paper-making subsidiary and Argos, the catalogue-shopping business. Racal Electronics plc, by selling some of the shares in its subsidiary Racal Telecommunications, established a market value for that substantial part of the Racal Group, since renamed Vodafone plc. Since all of these businesses were demerged as going concerns, the market values achieved no doubt exceeded the liquidation values of their assets.

The Bid Value. A variety of calculations can be carried out to judge how much to pay for a company as a going concern. The *Capitalised Average Earnings* are

$$\frac{\text{Earnings} + \text{Assumed Growth}}{\text{Required Real Return}}$$

This calculation assumes that the purchased company will be retained indefinitely, thus providing an infinite stream of earnings. The Earnings can be estimated from past performance. The Assumed Growth is the change that the purchaser believes can be achieved by improved management. The Required Real Return is the minimum return on investment that the purchaser is willing to accept, after taking inflation into account.

Alternatively, one can assume that the acquired company will be sold for some postulated price at some specific time after purchase. It is then possible to calculate the *Present Value of Future Worth*, the sum of the discounted price of the future sale and the capitalised earnings to the time of the sale. A few minutes' work with discount tables will produce this sum.

Yet another possibility is the use of appropriate *Exit Ratios*, those most often considered being:

$$\frac{\text{Market Price}}{\text{Earnings}} \quad \text{and} \quad \frac{\text{Market Price}}{\text{Cash Flow}}$$

The 'appropriate' values of these ratios are those applicable to businesses of a similar character that have recently changed hands. Suitable comparators may not be readily available, especially for a diversified group, but it may be possible to find values for the components of a group.

Example 6.2 Imperial Chemical Industries plc: Estimated Break-Up Value

---

The possibility that a bid may be made for a company's shares, or the appearance of an actual bid, triggers attempts by commentators to value the potential target. In this way we obtain well-informed estimates of break-up values, even if the contemplated bid does not materialise.

In May 1991 Hanson plc purchased in the market some 2.8 per cent of the shares of ICI, Britain's largest manufacturing company and the world's fourth largest chemical company. Hanson is a noted 'predator', which has bid successfully for large groups that were later broken up, in part at least.

It is hardly surprising that the purchase of this stake in ICI gave rise to speculation concerning the offer price that would be necessary to secure control.

Just prior to the Hanson purchase, ICI shares had traded at around £11 each. The price paid by Hanson was just below £12, and the market price then rose rapidly to about £13 per share. Commenting on the situation, *The Economist* magazine estimated the break-up value of ICI as indicated (in compressed form) in Table 6.2.

There is no obvious relationship between the profits of the individual business areas

Table 6.2   Estimated Market Value of Component Businesses of ICI in 1991 (values in £ million)

	Forecast Operating Profit	Estimated Market Value	Market Value / Oper. Profit
Pharmaceuticals	455	6000	13
Paints and surface-effect products	148	2050	14
General chemicals	145	860	6
Petrochemicals/Plastics/Fibres	5	1200	240
Explosives	45	400	9
Agrochemicals/Fertilizers	165	1750	11
Stakes in Related Companies	60	400	7
Totals	1023	12660	12

and the anticipated sale prices. Presumably each sector was valued in the light of its perceived long-term earning potential, rather than its performance in the recessionary conditions of 1991. Not surprisingly, in view of its profits and the generally high esteem in which pharmaceutical companies were held, that division makes a large contribution to the estimated value of the group.

At this time ICI had some 710 million shares in issue; hence the estimated break-up value is around £17.80 per share. We note that Hanson's initial purchase of a significant stake was at a price some 10 per cent higher than the then market level, that the market price subsequently rose to a level some 20 per cent above the pre-purchase level, and that the probable value of the components of ICI is some 60 per cent in excess of the market capitalisation before Hanson declared its interest.

It can safely be assumed that not all of ICI's shareholders think ill of Lord Hanson, the chairman of the company of that name. The Board of ICI did not receive his over-

PLATE 9

Bottles of ICI's 'Melinar', capable of complete recycling into other PET products.

tures with enthusiasm, however. It initiated a programme of 'reconstruction' which may involve demergers and the sale and closure of subsidiaries.

6.3 Share Transactions

Here we consider how shares come into the hands of investors, and how they change hands subsequently. These two processes are achieved through the *Primary and Secondary Markets* in shares. It is only at the primary stage that money passes from the investor to the company that issues the shares. Thereafter, it passes between successive owners of the shares. Fig. 6.1 shows some of the ways in which shares can change hands.

The Primary Market

When it wishes to issue shares, a company can offer them to the public, through an *Offer for Sale* or a *Public Issue*, or it can carry out a *Placing*, in which the shares pass to financial institutions in a pre-arranged manner.

The usual way for an already quoted company to issue new shares is through a *Rights Issue*. Here the shares are offered first to existing shareholders, at a stated price and in proportion to their shareholdings. Westland plc, for example, made a 2-for-5 rights issue in 1986 at a price of 60 p per share. This involved the allocation of 23.7 million shares and netted the company £14.3 million. The right to subscribe can be

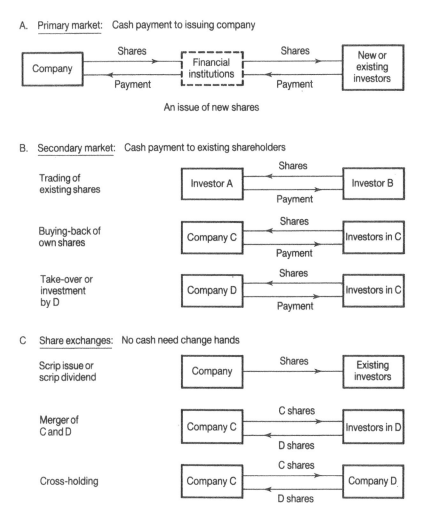

A. Primary market: Cash payment to issuing company

An issue of new shares

B. Secondary market: Cash payment to existing shareholders

Trading of existing shares

Buying-back of own shares

Take-over or investment by D

C Share exchanges: No cash need change hands

Scrip issue or scrip dividend

Merger of C and D

Cross-holding

Figure 6.1 Share transactions

sold, if a shareholder does not wish to 'take up his rights'. However, his proportion of the company's equity is then reduced, or diluted.

The price of the shares issued in this way is usually set below the current market price. This might be perceived to give the existing shareholders a special benefit, and to allow for possible dilution of earnings per share. In reality, the price is set with the aim of generating sufficient demand to absorb the additional shares. Unless the issuing company is highly esteemed, the effect of a rights issue is likely to be a prolonged depression of the market price of the shares.

The British Aerospace rights issue of October 1991 is instructive, though fortunately not typical. Before the announcement of the 2-for-5 issue at a price of 380 p, BAe shares had traded above 500 p. At the take-up date the shares were trading in the range 360 p to 370 p, and less than 5 per cent of the new shares were taken up. The rest remained in the hands of underwriters, financial institutions (some already shareholders) who had guaranteed to take the shares for a commission of 1¼ per cent.

The granting of options and the sale of warrants is in effect a deferred issue of new shares. Since the option will not be exercised until this is advantageous to the holder, these processes are somewhat like a rights issue.

With the authorisation of a general meeting of shareholders, directors can offer them the option of additional shares in the company in lieu of cash dividends. The acquisition price is generally linked to the market price at the time the dividend is declared. This *Share Distribution*, if it is available, together with the exercise of options, slowly increases the number of shares in issue.

With the authorisation of shareholders a company may *Buy Back* its own shares, either in part, to increase the degree of ownership of holders of the shares remaining in issue, or wholly, in order to 'take the company private'. Repurchase of shares is not common in the United Kingdom, but The General Electric Company plc had in 1990 the authority

> "to make market purchases on The Stock Exchange of up to an aggregate of 400 million Ordinary Shares of 5 p each, being approximately 14.8 per cent of the Company's issued share capital".

In the two years ending 31 March 1986 GEC actually purchased some 79.6 million of its own shares, representing about 3 per cent of its issued share capital.

Buying-back shares might appear to be an eccentric thing for a company to do, but the directors may argue that the best investment that they can identify is their own company. Certainly the act is an expression of confidence and should serve to support the price of the shares.

The Secondary Market

The transfer of ownership of already issued and paid-up shares has no direct financial effect on the issuing company. It may be of profound significance for the control of the company, however. A company's Articles of Association may restrict the ownership of shares, particularly voting shares; either beyond a certain fraction of those in issue, or by certain categories of owners, such as those resident abroad. The intention of such restrictions is obviously to influence the degree and nature of control that shareholders can exercise.

When a majority of the voting shares of a company come into one pair of hands, the consequences are considerable. The majority owners may seek to change the directors or senior managers of the company. In the limit, the company may experience a *Take-Over Bid*, an offer to all of its shareholders to buy their shares. This process is governed by a complex code of practice administered by the Panel on Takeovers and Mergers.

The price of its shares in the secondary market is important to a company and its employees for several reasons:

- It indicates how the company is perceived by the general investing community.
- It determines the cost of raising further share capital.
- It influences the behaviour of potential bidders for control of the company.
- It determines the value of the options held by directors and employees.

- It is a major constituent of the gains of shareholders from their investment in the shares, and companies are not averse to rewarding their shareholders.

What determines the market price of the shares of a particular company, or the general level of an entire sector of the market? The innumerable factors that determine the future development of the company or sector of the economy undoubtedly influence those buying and selling shares. At least as important is the psychology of the market itself. This was well understood by Lord Keynes, who made the following observations on professional investors:

> "They are concerned, not with what an investment is really worth to a man who buys it 'for keeps', but with what the market will value it at, under the influence of mass psychology, three months to a year hence."

He likened their thought-processes to those of competitors who seek to select from photographs of a hundred girls the six that will be most often selected by the whole field of competitors:

> "It is not the case of choosing those which, to the best of one's own judgement, are really the prettiest, nor even those which average opinion genuinely thinks the prettiest. We have reached the third degree where we devote our intelligences to anticipating what average opinion expects the average opinion to be. And there are some, I believe, who practise the fourth, fifth, and higher degrees."

King's College Cambridge benefited greatly from Keynes' insights; for a number of years he was responsible for its investments.

Share Exchanges

The processes considered above all involve the transfer of money, either to the issuing company or its shareholders. But some share transactions take place without money changing hands.

In Section 5.5 mention was made of a *Scrip Issue*, in which additional shares are issued without payment, to transfer funds from Share Premium Account to Share Capital. This process does not alter the par value of the shares. Much the same result can be achieved through a *Share Split*, in which the par value does change. For example, a 4-for-1 split might reduce the par value from 100 p to 25 p.

As was seen in Example 5.6, the *Merger* of two companies need not involve payment in cash by one company for the shares of the other. If the shareholders of the acquired company will accept the offer, the acquiring company can issue its shares to them in return for their shares in the acquired company.

Companies sometimes take *Cross-Holdings* in one another's shares. This might be done to indicate an intention to work together in the future, in a general fashion, rather than through joint ventures, but still without full merger. The creation of a cross-holding normally involves the exchange of agreed numbers of shares, although it can take place by purchases in the market.

Example 6.3 Northern Engineering Industries plc 1989,
Rolls-Royce plc 1989: Acquisition by Share Exchange

Northern Engineering Industries plc was acquired by Rolls-Royce, the aero-engine group, early in 1989. Its turnover was about one-quarter that of the acquiring company, and its businesses provided a significant diversification of the activities of the combined group. The NEI group contained several of the famous engineering concerns of the North of England, notably: Parsons, pioneering makers of steam turbines; Reyrolle, manufacturer of heavy electrical equipment; and International Combustion, one of the country's leading boiler companies. NEI's businesses fell into two broad classes: Energy Conversion (turbines, boilers, diesel engines, switch gear, transformers) and Materials Handling (cranes, mining and tunnelling equipment, structural steelwork).

The merger of the two groups took place through an agreed bid, with NEI shareholders offered seven RR shares for every ten NEI shares. In the event, Rolls-Royce obtained all the NEI ordinary shares, but it did not bid for the NEI preference shares. Hence a market in these shares remained, and NEI continued to report annually to its preference shareholders.

As the acquisition was achieved by an exchange of shares, Rolls-Royce did not have to find money to pay for this expansion of the group. (It had, however, earlier bought 4.7 per cent of the NEI shares in the market, at a cost of some £11 million.) Nevertheless, it is possible to establish the effective cost of the purchase. The price of a Roll-Royce share on the 17 April 1989 was 185 p. Hence each NEI share was effectively valued at $0.7 \times 185 = 129\frac{1}{2}$ p, and the 235 million ordinary shares in issue came to £304.3 million. Existing options add £6.5 million to this, and Rolls-Royce stated the cost to be approximately £310 million.

The purchase of 95.3 per cent of the NEI shares involved the issue of 158.3 million RR shares of par value 20 p. This increased the

Called-up Share Capital by £31.7 million. The Share Premium Account was not increased, since the sale was not for cash; indeed, this account was reduced by the share-issue costs of some £3 million. Instead, the premium of 165 p per share (£261 million in all) appeared in a Merger Reserve.

The Goodwill arising from the purchase amounted to £239 million, after certain 'fair-value adjustments' that removed £64 million from NEI's book value. The Goodwill was written off the Merger Reserve.

The net effects of these processes left Rolls-Royce owning all the ordinary NEI shares, and with an increase of about 20 per cent in the number of its own shares in issue. Its Shareholders' Funds changed little, from £949 million at the end of 1988 to £1001 million at the end of 1989 (disregarding £125 million of retained profit, which was not relevant to the transactions considered here). Hence the RR book value per share fell from £949/801.5 = 118 p to £1001/959.8 = 104 p as a consequence of the acquisition.

The table relates these values to others of relevance to the purchase. In this case there was only one bidder for ownership of NEI. Hence the bid premium was not as large as it might have been if there had been competition for the shares. Indeed, we see from the values in Table 6.3 that the effec-

Table 6.3   Share Prices and Values per Share

	NEI	Rolls-Royce
Market Prices		
End 1988	117 p	132 p
1st April 1988	132 p	183 p
1989 Range	142/113.5 p	131/202 p
End 1989	–	183 p
Book Values		
End 1988	51.5 p	118 p
End 1989	58.5 p	104 p
Bid Price	129.5 p	185 p

tive price paid by Rolls-Royce was somewhat lower than that reached in expectation, prior to the bid.

In Section 6.2 a number of ways of estimating the appropriate bid price were suggested. For NEI the Capitalised Value of Earnings is found to lie in the range 50 p to 95 p, for required returns of 10 and 15 per cent. The Present Value of Future Worth is found to be between 85 p and 115 p, for discount rates of 10 and 15 per cent, and postulating the sale of NEI to occur either

10 or 15 years in the future, at the effective price paid by RR. Plausible Exit Ratios (13 for Price/Earnings and 5 for Price/Cash-Flow) lead to values around 110 p to 115 p.

Evidently Rolls-Royce perceived value in NEI beyond that indicated by any of these calculations. Perhaps a marked improvement in performance was anticipated, or benefits from 'synergy' in the product lines or Research and Development programmes of the merged companies.

6.4 Other Traded Securities

Government Stock

The securities issued by governments affect companies in several ways. By going into debt themselves, governments enter into competition with companies for the finite funds available for investment, so that the general cost of capital is increased. Moreover, by changing the interest rate on the bonds or other securities that it issues, the government influences the interest on company securities. The interest rate for company fixed-interest securities is generally a little above that offered by the government, since companies are more likely to default on their debts.

On the other hand, government securities are of service to companies, by providing a secure interest-earning store within which surplus cash can be deposited. As these securities are traded on stock markets, the invested funds can readily be recovered, subject to changes in market prices.

The stocks sold by the UK government are called *Gilt-Edged Stock* or simply *Gilts*; they are issued in units of £100 nominal value. These stocks represent the National Debt and are issued when the government spends more than it receives. Net redemptions can occur when government revenues exceed its expenditure. Rather similar stock is issued by local authorities, by agencies such as nationalised industries, and by overseas governments and their agencies.

On issue, these securities have terms in the range, say, five to twenty years. During that period their market prices vary, driven by changes in prevailing interest rates and occasionally by concern about the certainty of redemption.

Fixed-interest securities, in particular those of overseas governments, are sometimes called *Bonds*.

In addition to fixed-interest stock, the UK government issues *Index-Linked Gilts*, whose redemption value and interest are increased in line with an index of consumer-price inflation in Britain. While these are of use to companies that have long-term, predictable commitments, notably pension providers, they are of little interest to manufacturing companies.

Another kind of government security is the *Treasury Bill*, which commonly has a term of ninety-one days. These bills carry no interest, but are issued at a discount to the redemption price. Once issued they can be traded, and this trading provides securities of even shorter term.

While the market price of loan stock, bonds and gilts is influenced by the interest that they offer, the time to redemption also affects the price. For any given security one can distinguish the *Interest or Flat Yield*, which is the annual interest payment divided by the current price, and the *Redemption Yield*, which is an effective value incorporating the repayment of the nominal value on redemption.

Loan Capital

The fixed-interest securities issued by companies possess many of the features of government stock. Like shares, the various classes of *Loan Stocks* may be traded. As in the case of shares, there exist primary and secondary markets in these securities. Since the interest payments are fixed, the prices of loan stocks do not fluctuate as widely as do those of ordinary shares. This statement assumes that the issuing company is judged able to continue paying interest and ultimately to repay the debt when it falls due.

Loan stock is commonly issued in units of £100 nominal value. It is usually *Redeemable*, that is, the nominal value is repayable at some point. The *Coupon Rate of Interest* is relative to the par value, but the effective rate changes as the market price varies. While the interest rate is usually fixed, variable-rate securities may be issued in periods when interest rates are volatile.

Each issue of loan stock is assigned a *Priority* in repayment in the event of the issuing company being wound up. That with first claim is termed *Senior* stock, and those further down the line are more *Junior*. Stock with second call upon residual funds may also be termed *Subordinated*. The interest on junior issues is somewhat higher than that on senior stock, to allow for the higher risk.

Some loan stocks are convertible to ordinary shares. They differ from convertible preference shares in having, before conversion, the prior claims that are characteristic of debt.

Example 6.4 BICC plc 1989: Market Value of Loan Stocks

BICC (formerly British Insulated Calendar's Cables) is a major supplier of cables, for both power and information transmission, in many parts of the world. Its subsidiary Balfour Beatty Ltd is a significant company in construction and construction-related engineering. Other subsidiaries based in Australia, the United States and Canada give the group a strong overseas presence.

The borrowings of the group reflect the international spread of its activities, being denominated not only in sterling, but in Australian, US and Canadian dollars, Italian lire and other currencies.

Table 6.4 shows some of the fixed-interest borrowings of the group; preference shares are included, since they are in many ways similar to debt. The spans of years indi-

Table 6.4 Fixed-Interest Securities

	Indicative Market Price
7% debenture stock 85/90	£96.40
7¾% debenture stock 90/95	£88.50
7½% US$ subordinated debenture stock 2011	$83.30
4.2% first cumulative preference stock units of £1	38.2 p
3.85% second cumulative preference stock units of £1	35.0 p

cated are those within which redemption may occur. The market prices will vary continually, but an attempt has been made to indicate the levels that might have been expected in 1989. For the purposes of these estimates, it has been assumed that the ruling interest rate on long-term securities of sound commercial organisations was 11 per cent in the UK and 9 per cent in the US.

The year 1989 is particularly significant for the first security on the list, which was due for redemption a year later. Since an investor seeks a return of 11 per cent to redemption, the price he is willing to pay is

$$£100 \times (1 + I)/1.11$$

where I is the coupon interest. For $I = 0.07$, we obtain the price of £96.40.

For a security that is either irredeemable, or will be redeemed only in the distant future, it is the stream of interest payments that determines the price. Hence the investor seeking a return of 11 per cent on his purchase price is willing to pay

$$£100 \times I /11$$

More complicated formulae apply for intermediate cases, but they need not concern us here.

We note that the price of these fixed-interest securities depends upon:

- the ratio of the prevailing interest rate to the coupon rate, applied to the nominal value
- the country in which the transaction takes place, since prevailing rates vary from country to country
- the time to redemption, and the conditions under which it can take place, and
- the repute of the issuing company, that is to say investors' perception of its ability to pay interest and the capital sum.

Capital Structure: Equity or Loans?

The first question to be asked is: 'How do companies decide whether to raise capital by borrowing or by issuing shares?' Obviously they will be influenced by what they have to pay for money obtained in these two ways. Much of this chapter is therefore concerned with determining of the cost of capital.

7.1 Influences on the Choice

Let us try to put ourselves in the place of a Finance Director who must recommend to his board of directors a way of raising additional capital. Some of the thoughts passing through his mind are suggested below.

   Have recent calls for capital by other companies been for loan or share capital? (Potential buyers could be sated, so it might be better to do the opposite.) What kind of capital did we raise most recently ourselves? (Could be a good idea to change tack.) What kind of 'paper' are our present investors likely to prefer? (I could phone around and ask.) What balance between equity and debt is typical in this country, or in this industry? (You never get anywhere by trying to be different.) Is the proposed use of the new capital dynamic enough to attract equity investment? (Paying down short-term loans is not very exciting.)

Are current interest rates high by historical standards? (If so, raising debt does not look too good an idea. Maybe we should look overseas, where rates are lower.) Is inflation high by historical standards? (If so, maybe debt is a good idea, since its real value will be eroded quickly.) Is inflation likely to fall shortly? (Better not get tied into high interest rates now.)

Is the current share price relatively high? (If so, going for equity looks the cheaper solution, though a rights issue would knock back the price.) Are the identifiable assets sufficient to cover the current share price? (If so, providers of either equity or debt should queue up to take whatever we issue.)

Are our profits sufficiently larger than the present interest payments to carry more debt? (If not, better steer clear of it.) What is our tax position? (Need to bear in mind the extra tax on overseas earnings, if our dividend pay-out becomes too large.) Are we growing fast enough to maintain the dividend and to increase earnings per share on a bigger share base? (If not, a share issue looks dangerous.) If we issue more shares, is a predator likely to gain control of the company? (If so, debt seems the answer. That may limit our growth, though.)

After all these thoughts have passed through his mind, our Finance Director may feel that neither equity nor debt meets his needs. One of the hybrid securities may seem preferable. Table 7.1 indicates some of the options open to him. The interest and dividend rates indicated there (loans a little more expensive than debentures, convertible and preference stock a little cheaper, and ordinary dividends still lower) are those pertaining to securities issued at the same time. Subsequently, very different relationships may develop as interest rates change and the ordinary dividend is increased.

Table 7.1 shows also some advantages and disadvantages of the various forms of capital, as seen by the company and by providers of capital. The tax implications are extremely complex; a few of them will be explored in Chapter 8.

7.2 Cost of Capital

One important question was not asked explicitly by the Finance Director whose thoughts we have imagined: 'What combination of capital-raising securities will minimise the overall cost of capital to the company?' In fact, some of the particular questions he did ask himself bear upon this fundamental issue. It may not be possible to provide a completely convincing answer, but a consideration of the problem gives useful insights into the roles of debt and equity capital.

Weighted Cost of Capital

In seeking to minimise the cost of capital, we must first define the quantity we hope to control. There are other reasons for needing to know the average cost of capital: to discover whether in this respect the company's performance is improving or declining, and to decide whether the return on a proposed investment is sufficient to justify raising capital to fund it.

To determine what a company pays for its capital, we create a weighted average recognising all the securities currently in issue, both borrowings and equity. It is easy

Table 7.1   Typical Characteristics of Company Securities

Type	Benefits to Provider of Capital	Control of Provider of Capital
Debentures	Fixed Interest (FI). Usually redeemed in time. Secured against an asset.	None unless company defaults on payments.
Loans	Fixed Interest (FI +). Redeemable or recallable on demand. Priority on liquidation.	As above. Can refuse to renew a loan.
Convertible Shares	Fixed Interest (FI −). May be worth converting. Sometimes redeemable.	Class vote on major issues. Conversion to Ordinary.
Preference Shares	Fixed Dividend (FI −). Possibly cumulative. Sometimes redeemable. Priority over Ordinary.	Class vote on major issues.
Ordinary Shares	Variable Dividend (FI − −). Seldom redeemed. Market price may rise or fall significantly. Usually offer liquidity. Potential protection against inflation.	Elect directors. Vote on major and routine matters. Limited by influence of directors and managers. Limited in default.

Type	Risk to Provider of Capital	Risk to Company
Debentures	Erosion by inflation.	First charge on income. On default: receivers, liquidation
Loans	Erosion by inflation.	First charge on income. On default: receivers, liquidation
Convertible Shares	Unpredictability of price of Ordinary shares.	Priority call on income before conversion. Dilution after conversion.
Preference Shares	Default. Erosion by inflation.	Priority call on income. Advance Corporation Tax on dividends.
Ordinary Shares	Cut in dividend. Loss of capital. Variability in share price. Possibly contrary interests of directors.	Dividend outflow modest. Dilution of existing holdings. Reduction in control. Advance Corporation Tax on dividends.

to determine the *Cost of Debt*, as the average interest rate for the Loan Capital:

$$R_d = 100 \times \frac{\text{Total Interest on Loan Capital}}{\text{Loan Capital}} \qquad (7.1)$$

The *Cost of Equity* is a more difficult concept, which will be explored in Section 7.3. For the time being, we simply denote it by the symbol R_e, expressing the cost of equity as a percentage. We can now calculate the cost of all the company's capital as the weighted average

$$R = W_d R_d + W_e R_e \qquad (7.2)$$

Here W_d is the fraction of the capital made up by borrowings, and W_e is the fraction provided by equity, so that

$$W_d + W_e = 1$$

relates the two fractions.

Gearing and Risk

The ratio of loan capital to equity, W_d / W_e, is referred to as the *Gearing* of the company's capital. The usual American word for this quantity, *Leverage*, conveys much the same idea. For practical application a more precise definition is needed. We need to decide, for instance, where to put preference share capital. In Section 9.6 more precise definitions of gearing will be introduced.

The problem under investigation can now be expressed: 'How does the average cost of capital vary with gearing?' To answer this question we look first at the reason for the dependence on gearing of the expected returns on both kinds of capital. This requires that we consider the *Risk* that investors perceive when they buy particular kinds of securities.

In discussions of investment the word Risk is usually taken to mean simply variability, rather than a more varied spectrum of undesirable consequences. Thus a 'risky' investment is one whose outcome is uncertain, offering the possibility of rather large as well as rather small benefits. Looking at Table 7.1, we observe that ordinary shares offer both the greatest number of potential benefits and the greatest variety of drawbacks. Even when cataclysms are avoided, the outcome from investment in shares is strongly variable and thus impossible to predict with great confidence.

In seeking to express these ideas more precisely, we consider first the relationship between the costs of the two kinds of capital. In doing this, we must recognise the reciprocal relationship between cost of capital and return upon investment: the cost to a company is the return to investors. It appears that typical investors are *Risk-Averse*: they seek investments that appear to offer the minimum risk or variability for a given return. Alternatively, typical investors expect higher returns from investments that appear to be more risky. (There are, of course, some inveterate gamblers who are attracted by risk.) In terms of the symbols introduced above, this implies that the cost of equity exceeds the cost of debt:

$$R_e > R_d \tag{7.3}$$

for any particular gearing.

The elevation in the *Expected Return* on equity above that on debt is known as the *Risk Premium*. Similar premiums exist within each class of securities; company loans, for instance, offer higher interest than do government stocks, since there is a greater chance that a company may default on its debts.

The next step in the argument depends on the observation that gearing is itself a measure of risk. Providers of equity and providers of debt are both likely to perceive a highly geared company as more risky. From whatever trading profit it may make, a large interest charge is deducted before the profit is struck. Hence the providers of equity capital will perceive a risk that there will be insufficient profit to justify paying their dividends. And both kinds of investors will fear that a downturn in trading (or a rise in the interest charged on variable-rate debt) will bring the company into default through inability to service all of its debts. Thus we conclude that all risk-averse investors will require higher levels of reward (that is, higher values of R_e and R_d) as the gearing increases.

The arguments advanced in the preceding paragraph do not have much weight for low gearing, that is, for a company that has a level of debt that can easily be carried, even in bad times. Hence the expected returns for moderate levels of debt will be only a little greater than those for no debt at all.

Fig. 7.1 presents variations of the two returns upon capital (the solid lines) that exhibit the behaviour that has been deduced. To provide a measure of indebtedness that varies between 0 and 100 per cent, we show the variation against the debt fraction W_d, as well as against gearing W_d/W_e. Between the two solid curves lies a dashed curve representing the average or weighted cost of capital, given by Equation (7.2). Subject to the validity of the plausible arguments set out above, this rather flat curve will exhibit a minimum value for a moderate level of indebtedness.

Interpretation

In Fig. 7.1 the cost of debt for very high gearing ($W_d \simeq 1$) is shown as a little greater than that of equity for low gearing ($W_d \simeq 0$). It is instructive to consider why this should be so. Loan capital is normally composed of tranches of debt with different levels of security and priority of repayment. Hence for a company with no equity at all, the junior, most subordinate, least well-secured debt plays the role of equity. It provides the senior debt with a buffer against adversity. Since the junior debt does not have the potential to share in unexpectedly large profits (unless it is convertible), those who hold it will require a return above that expected by holders of equity in an ungeared company.

The role of equity as a cushion was explicit in the capital structure adopted by Eurotunnel plc, the company set up to construct and operate the Channel Tunnel. Most of the capital was in the form of debt, but the lending banks insisted that there

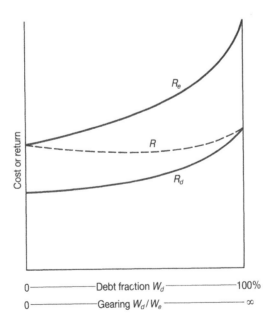

Figure 7.1   Average cost of capital R derived from costs of (or expected returns on) debt R_d and equity R_e. W_d and W_e are the contribution to capital from equity and from debt

should be a tranche of risk-absorbing equity, in case the initial projections proved too optimistic. In the event, the costs of digging and outfitting the tunnel were higher than the £4.8 billion envisaged in 1987; inflation above that expected played a part. Late in 1990, about the time the tunnels met under the Channel, a further £2.7 billion in capital had to be raised. Again this was mostly in the form of loans, but a rights issue ensured that the bankers retained the protection they required.

Before moving on, we should note that the arguments advanced above are not universally accepted. An alternative analysis starts from the premise that the average cost of capital should be independent of gearing; the effect of this postulate is to change the flat dashed curve of Fig. 7.1 into a straight line. Strange though it may seem, what appear to be fundamentally different models are difficult to test empirically. What with trade cycles, fluctuations in interest and exchange rates, and variations in the fortunes of particular companies, it is not possible to determine the variations we have been discussing with sufficient precision to decide between these rival theories.

Perhaps it does not matter much which model is 'correct', since both suggest that the average cost of capital will be nearly constant over a considerable range of debt and gearing. Whatever the theorists may say, companies and investors behave as though there is a relationship of the kind shown by the dashed line in Fig. 7.1. It is not unusual for established companies to operate with gearing in the range 30 to 60 per cent for years on end. When debt strays above this level, possibly following the acquisition of a large subsidiary, analysts are apt to suggest that some corrective action is appropriate, say the sale of assets, or a rights issue. If debt falls below the conventional range, the same observers are likely to suggest (save in hard times) that the directors have lost direction, are behaving too conservatively, and are not making good use of their shareholders' capital.

To complete this discussion of the level of debt that companies choose to carry, we note that the conventional gearing ratio will often overestimate the contribution made by debt to a company's total capital. For a group that has over the years acquired many subsidiaries, and has thus written off substantial amounts of Goodwill, the book value of equity represents only a fraction of the funds subscribed by the shareholders. Hence the ratio Debt/SHF combines one firmly established number, Debt, with a value that is usually an underestimate of the actual Equity element of a company's capital.

Example 7.1 Thorn EMI plc 1986/90: Balance of Debt and Equity Capital

---

Thorn EMI was created by the merger of Thorn, a company whose roots were in lighting, and EMI, whose foundations were in recorded music. Currently, the group's largest and most profitable business sector is Rental and Retail, mostly of electrical equipment for communications, domestic use and entertainment. Nearly as large is its Music activity, the recording of popular and classical music and the manufacture and distribution of the media on which they are recorded. At one time the group intended to sell its Lighting business, but it then decided not to do so, and has made a number of acquisitions in that area. The other businesses of the group are described as Technology. Their diverse products include security systems, software, and electronic ticketing systems.

The group's activities changed significantly over the five-year period to be considered. At the beginning of this period, Thorn EMI made major domestic appliances (business sold to Electrolux of Sweden), television sets (Ferguson, sold to Thomson of France), semiconductors (INMOS, sold to the Italian-French SGS-Thomson), meters (business sold to the American Schlumberger) and kitchen appliances (Kenwood, gone in a management buy-out). During the period it expanded its Music and Rental and Retail sectors.

Table 7.2 shows the response of the group's capital structure as this refocusing took place. The upper part indicates how the Capital Employed by the group was made up. We note that the Equity remained fairly constant during the period, while the Borrowings varied significantly. The Capital Employed would also have remained fairly constant, save for the growth of Minority Interests.

The central part of the table gives some of the main flows of funds into and out of the group and the sums written off as Goodwill. In the four years from 1987 to 1990 a total of £516 million was generated by sales of shares and warrants, while £335 million came from retained profits. Why then did the SHF not reflect this? The answer lies in the Goodwill written off against reserves, amounting to £637 million in these years.

This part of the table demonstrates the role of acquisitions in reducing the book value of the equity below the total of the funds introduced by shareholders. One sees also how the sale of assets first reduced the group's debt, and then in part funded the acquisitions that cost £988 million over the years 1987 to 1990.

At the bottom of the table are given the two ratios conventionally used to characterise a company's capital structure; it is the first (Debt/SHF) that is usually referred to as the gearing. After a fall in indebtedness while the group was focused by the sale of businesses, debt rose again until the gearing attained the level around 50 per cent that the company apparently sees as appropriate in normal operating conditions.

Table 7.2   Changes in Capital Structure 1986/1990

	1986	1987	1988	1989	1990
Equity	581	630	645	590	643
Net Borrowing	337	240	58	167	316
Provisions	115	110	132	170	172
Minority Interests	23	28	126	229	233
Capital Employed	1056	1008	961	1156	1364
Issues of Shares, etc.	6	14	382	108	6
Retained Profit	(61)	32	59	96	148
Reduction in Net Debt	19	98	181	(108)	(149)
Sales of Assets	179	157	250	213	265
Acquisitions	–	(18)	(481)	(205)	(284)
Goodwill	–	(15)	(396)	(166)	(98)
Net Debt/Equity	58%	38%	9%	28%	49%
Net Debt/Capital Employed	32%	24%	6%	14%	23%

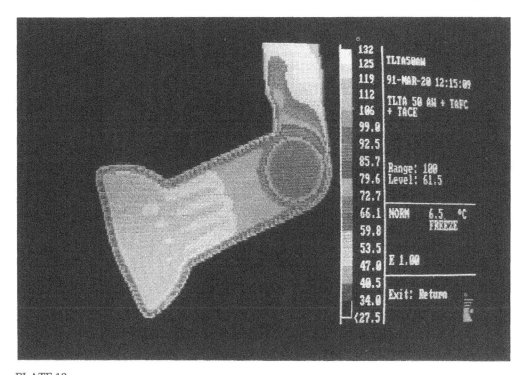

PLATE 10

Thermal testing of an ARIA spotlight made by Thorn Lighting.

7.3 Cost of Equity

The Problem

We seek an 'interest rate' that represents a company's reward to its shareholders for the equity they have provided. The capital that is of interest is measured by the Shareholders' Funds, comprising both the Share Capital that was positively invested and Reserves built up over the years in the name of the shareholders. Since all shareholders, old and new, are treated equally, the return to new shareholdings created by issuing additional shares is also the cost of that capital to the company, that is, to the present shareholders.

An obvious return to shareholders is the dividend actually paid to them. It is the *Gross Dividend* that measures the cost to the company. This comprises not only the *Net Dividend* sent to the shareholder, but also the notional *Advance Corporation Tax* on that dividend, which is retained by the company and remitted to the tax authorities.

The shareholder benefits also from the growth of the company, through investment of the retained element of the annual profits and through, for example, increases in the value of property owned by the company. Hence it is not just the dividend, but some measure of the increment in the value of the company that determines the total return to the shareholder and the total cost to the company.

An additional complication is the need to predict the future development and *Dividend Policy* of the company. When a long-term loan is taken out at a fixed interest rate, one knows the nominal payments throughout its life, though the real cost after inflation is somewhat uncertain. There is no certainty as to the dividends that will be paid by a company; they are determined by a policy conceived by the directors and subject to their changing perceptions of the needs of the company. There is even less certainty as to the retained profits and other additions to reserves.

Finally, it is necessary to allow for the different ways in which company tax is levied on interest and on equity costs. Interest payments are accepted as expenses and reduce the taxable income of a company. (However, most recipients of interest presumably pay tax on it, and this will influence the level of interest they require.) A company's profits after interest are normally taxable; in particular, the *Advance Corporation Tax* associated with dividends is paid quite quickly. Hence equity capital costs a company more than a superficial look at the profit and loss statement may suggest. It is sometimes argued that the capital structure of companies is biased towards debt by these differences in taxation.

Some Possible Solutions

We shall consider several ways of determining the cost of share capital, seeking to allow for the complexities mentioned in the preceding paragraphs. For the purpose of this discussion we take the gross dividend to be 5 per cent, and the interest on debt to be 10 per cent. While these values are chosen for illustrative purposes, they are realistic during certain phases of the economic cycle and for particular industrial sectors.

The Risk Premium. It was argued in Section 7.2 that shareholders require a higher return on equity, to compensate for the additional risk inherent in it. Having adopting a 10 per cent pretax return on debt, we are led to anticipate that the total pretax return to shareholders must be around 15 per cent. For a gross dividend of 5 per cent, the component of return retained within the company must then be about 10 per cent, that is to say, around twice the more obvious return through the dividend. Expressed in symbols, these propositions give

$$R_e = R_d + RP \quad \text{(BT)} \tag{7.4}$$
$$= 10 + 5 = 15 \text{ per cent}$$

where R_e and R_d are, as in Section 7.2, the costs of equity and debt, and RP is the risk premium. The label BT identifies the return as being taxable; the label AT will be used later to identify returns that have already been taxed.

This line of argument can be expressed more precisely by distinguishing the different levels of risk associated with specific companies or industries. The risk premium will be measured from the base level *FI*, the interest available on an ideal risk-free investment, to which government securities provide the closest approximation. The risk premium itself is related to the base level, through

$$RP = \beta \, (RM - FI) \tag{7.5}$$

where *RM* is the average return for companies listed by the stock exchange. This can be found by combining average dividends with changes in an index of the overall behaviour of share prices. The difference (*RM* − *FI*) is the risk premium for the stock market as a whole. The Greek letter β, *Beta*, is a measure of the risk of an individual share issue relative to that for the entire market. That is to say, it indicates whether the share price fluctuates more (β > 1) or less (β < 1) than the market index.

Putting the pieces of this argument together, we can express the cost of equity as

$$R_e = FI + β (RM − FI) \quad \text{(BT)} \tag{7.6}$$

For β = 1, the value for the stock market as a whole, this formula gives $R_e = RM$, as would be expected.

Let us insert some plausible numbers. We take the fixed interest to be *FI* = 9 per cent, a little lower than the value R_d = 10 per cent that companies have been assumed to pay on their borrowings. Over a period of some decades the average market return has been *RM* = 14 per cent or thereabouts, a little less than the estimate of Equation (7.4). For the engineering companies of greatest interest here, the returns are more variable than those for the entire range of companies with quoted shares, which includes rather stable activities such as brewing and food retailing. For the engineering sector the index of variability Beta is around β = 1.3. Equation (7.6) then suggests a cost of equity of

$$R_e = 9 + 6.5 = 15.5 \text{ per cent}$$

for engineering companies.

It is sometimes said that: 'Equity capital is expensive capital'. The estimates obtained above provide some justification for this view. It must be remembered, however, that equity capital imposes a smaller cash drain upon the company, and that inability to pay a dividend does not have the dire consequences of default on a loan.

An instructive example of the effect of investors' expectations and perceptions of risk on the required rate of return is provided by the electricity-generation sector of the economy of the United Kingdom. As a state monopoly, the Central Electricity Generating Board was required by the government to achieve only a low rate of return upon investment, mainly because of the low risk arising from its monopoly position. In the competitive environment following privatisation the generating companies were expected to provide a higher return upon new investment.

Past Performance. Another approach to the cost of equity is through financial results reported in earlier years. A specific company can be assessed in this way, or one of its business segments, or the entire industrial sector of which it is a part. Several measures of a company's performance might be selected.

We start with Earnings, which incorporates both retained profit and dividends. This quantity can be converted into the equivalent of an interest rate by dividing the earnings per share by the market price per share:

$$\text{Earnings Yield} = \text{EPS/Price} \quad \text{(AT)} \tag{7.7}$$

This ratio has the disadvantage of changing as the share price fluctuates, but an average value can be chosen. In its favour, we note that the market price is not an arbitrary accounting quantity, but expresses the cumulative judgement of numerous investors.

Another way of characterising earnings is through its relationship to shareholders' funds:

$$\text{Return on Equity} = \text{Earnings/SHF} \quad \text{(AT)} \tag{7.8}$$

Although this ratio is commonly called Return on Equity, it may not be the 'cost of equity' that we hope to find. We have seen (Sections 5.6 and 6.2) that SHF is a somewhat arbitrary quantity, strongly influenced by a company's policies on writing-off and revaluation.

We can also relate earnings to quantities defining a company's capital, and specifically to net assets, which is equal to capital employed:

$$\text{Return on Investment} = \text{Earnings/Net Assets}$$

$$= \text{Earnings/Capital Employed} \quad \text{(AT)} \tag{7.9}$$

This divisor is somewhat better defined, but it may seem perverse to base the return on equity on the sum of equity and loan capital.

In using any of the proposals of Equations (7.7) to (7.9), we must allow for the tax that is entered in the profit and loss statement before the earnings are struck. This suggests that, for typical rates of taxation, the cost of equity may be some 40 per cent higher than the values given by these equations.

Finally, we might combine the current dividend, which is received immediately, and the anticipated future growth of the company, which can be estimated from past performance. Summing these components, we have

$$\text{Total Return} = \text{Gross Dividend} + \text{Growth Factor} \quad \text{(BT)} \tag{7.10}$$

The present dividend is known, but the second term is less certain. Should it be based on growth in earnings? In turnover? In assets? What period of time should be considered? How do we allow for that part of the growth that comes from the injection of additional capital? For a company with reasonably stable capital and capital structure, the growth in EPS over, say, the past decade might be taken to represent its performance.

Market Return. The return on an investment that is seen by the individual investor does not depend, at least in the short term, on the prior performance of the company or on long-term trends in the cost of capital. The investor is interested in

$$\text{Market Return} = \text{Net Dividend Paid} + \text{Taxed Capital Gain} \quad \text{(AT)} \tag{7.11}$$

for the coming year, or perhaps an even shorter time. Bear in mind the observations of Lord Keynes, quoted in Section 6.3.

Over a considerable period this return may coincide with the cost of equity as seen by the company, allowance being made for corporation, income and capital-gains taxes. On a shorter time-scale the return to the investor is dominated by the precise time of share purchase and, if the return is to be realised, the time of sale. For advice on the timing of share transactions, the investor can do no better than to draw upon the wisdom of J. Pierpont Morgan. When asked to predict how the stock market would move, that noted financier replied: "The one certain truth about markets is that they fluctuate".

Section 9.9 considers in more detail ways in which investors can assess potential equity investments.

Example 7.2 Results for the Period 1983/87

This period has been chosen to illustrate the proposals developed above, since its unusual features give the results special interest. This was a period in which:

- Share prices rose very strongly, until October 1987, when they fell abruptly.
- The 'productivity' of British industry increased greatly, that is, many companies produced more with fewer employees.
- Leading manufacturing companies developed important aspects of their operations overseas.
- Domestic inflation rose as high as 5.7 per cent and averaged perhaps 4.5 per cent
- There was steady growth in the whole UK economy, while manufacturing industry grew less strongly.

By considering the performance of a number of well-established companies, one can form a picture of the development of British engineering in the period. Looking at companies ranging from Ford UK at the large end to Ransomes, a manufacturer of grass-cutting machinery, at the other extreme, we find that typically:

- Turnover increased by about 8 per cent per annum, some 3.5 per cent above the rate of domestic inflation – although this comparison is not entirely relevant, since on average 60 per cent of the sales of these companies took place overseas.
- Earnings per share rose at around 30 per cent per annum, from the very low levels of the early 1980s.
- The number of employees fell by some 4 per cent per annum, so that Sales/Employee rose at an annual rate of about 12 per cent.
- The Earnings Yield defined in Equation (7.7) climbed from about 6.5% on 1 September 1987 to 10.5% on 1 December 1987, reflecting the rapid fall in share prices over that short period.
- The ratio Market Price/Net Assets fell

from about 1.5 on 1 September 1987 to 1.1 on 1 December 1987.

What do these figures tell us about the Cost of Equity to these companies? Applying a taxation correction of 40 per cent to the Earnings Yield, we have

R_e = 9 per cent using the 1st September yield
= 14.5 per cent using the 1st December yield

The first value is significantly lower than the estimates obtained in other ways. This suggests, as investors concluded in 1987, that shares were over-valued in the summer of that year.

The Return on Investment of Equation (7.9) can be expressed as

Earnings Yield × Market Price/Net Assets
= 11 per cent using values for either date

With a 40% uplift to allow for taxation, this suggests that R_e = 15 per cent.

For these companies SHF typically comprises about 65 per cent of the Net Assets. The Return on Equity of Equation (7.8) can then be written as

(Earnings/Net Assets) × (Net Assets/SHF)
= 11 / 0.65 = 17 per cent

Again applying the 40 per cent uplift, we obtain the suggested value R_e = 23 per cent. As might be expected, this is much higher than the other estimates, since SHF is not a very satisfactory measure of the true shareholders' equity in a company. Surprisingly, the Return on Assets provides a more plausible estimate: $11 \times 1.4 \simeq 15$ per cent. Seemingly, the inclusion of loan capital in the divisor compensates roughly for the depression of shareholders' funds.

Finally, we consider the growth model of Equation (7.10). Using a gross dividend of 5 per cent and the rate of growth in Turnover, we obtain

$$R_e = 5 + 8 = 13 \text{ per cent}$$

The increase in EPS was uncharacteristically high in this period, and quite obviously unsustainable; the return based on that factor can be set aside as an aberration. However, the rate of growth in Productivity gives

$$R_e = 5 + 12 = 17 \text{ per cent}$$

This is somewhat higher than most of our estimates, reflecting the impossibility of sustaining the rate of manpower reduction that has been incorporated in the change in productivity.

The more believable of the estimates found here lie in the range 13 to 17 per cent. Their average of 15.3 per cent falls between the values suggested by general arguments relating to risk, expressed in Equations (7.4) and (7.6). Taken together, these varied arguments point to a cost of equity around 15 per cent for engineering companies.

7.4 Concluding Words on Debt

We are now in a position to summarise the features of debt that induce companies to include loans as well as equity in their capital base. Debt has the following desirable attributes:

- It provides a hedge against inflation.
- It does not dilute the control of existing shareholders over a company.
- It does not dilute the earnings of the company indefinitely into the future.
- In the long term it usually proves cheaper than equity, in part because of the different tax treatment given to the two kinds of returns to providers of capital.
- It gives providers of equity capital a share in the earnings generated by a larger capital base.
- It offers the possibility of locking into low-cost capital, through selection of an appropriate time to issue debt.
- It taps funds that investors will make available only at fixed interest and with defined repayment.
- It can often be raised relatively quickly and confidentially and with low transaction costs.

Balancing these considerable advantages is the increase in risk that results from the inclusion of loan capital in the balance sheet and the appearance of associated interest payments in the profit and loss statement. One aspect of this risk has been noted above, namely, the possibility of enhanced earnings from the expanded capital base. The following example illustrates the corresponding down-side risk.

Example 7.3 Effect of Gearing on Profits and Retentions

We consider three capital structures, each involving the same total capital, taken to be 10,000 units. In each case it will be assumed that dividends are paid at 5 per cent on the equity component of capital, even when there is insufficient profit to cover the dividend payment. Interest and tax are levied at the same rates in each situation.

In order to investigate the dependence of earnings on the level of debt carried, we consider a range of Profits BIT (before interest and tax). One of the companies considered carries no debt at all. The second raises half of its capital in the form of debt, giving a conventional gearing of Debt/Equity = 100 per cent and Debt/Capital = 50 per cent. The third company raises most of its capital as debt, with a conventional gearing of 400 per cent and Debt/Capital = 80 per cent.

Tables 7.3, 7.4 and 7.5 show how Earnings and Retained Profits respond as the Trading Profit changes, as it might during the course of an economic cycle. The range of profits considered is by no means extreme: it omits cases in which the basic profit vanishes completely or becomes a loss.

The final four rows of each table present a number of relationships between the quantities determined in the body of the tables. We note that:

- In the most optimistic of all the scenarios considered here, the company retains only half of the Profit *BIT*. In less happy trading conditions the combination of interest, tax and dividends draws more funds from the company than are provided by its trading.
- In the absence of debt the company remains profitable over a wide range of trading conditions, and for a considerable range of interest rates. Even after paying a dividend it is usually able to retain part of the profits. However, the Return upon Equity is modest (on the assumptions made here, between 4 to 18 per cent), even in the most buoyant conditions envisaged.
- For gearing of 100 per cent the variability

Table 7.3 Performance of Company without Debt (Capital = Equity = 10,000)

Profit BIT	500	1000	1500	2000	2500
Interest (15%)	–	–	–	–	–
Profit BT	500	1000	1500	2000	2500
Tax (30%)	(150)	(300)	(450)	(600)	(750)
Profit AT (Earnings)	350	700	1050	1400	1750
Dividend (5%)	(500)	(500)	(500)	(500)	(500)
Retained Profit	(150)	200	550	900	1250
Retained/Profit BIT	−30%	20%	37%	45%	50%
Earnings/Equity	4%	7%	11%	14%	18%
Retained/Equity	−1%	2%	6%	9%	13%
Range			——— 14% ———		

Table 7.4   Performance of Company with Gearing of 100 per cent (Equity =Debt = 5000)

Profit BIT	500	1000	1500	2000	2500
Interest (15%)	(750)	(750)	(750)	(750)	(750)
Profit BT	(250)	250	750	1250	1750
Tax (30%)	0	(75)	(225)	(375)	(525)
Profit AT (Earnings)	(250)	175	525	875	1225
Dividend (5%)	(250)	(250)	(250)	(250)	(250)
Retained Profit	(500)	(75)	275	625	975
Retained/Profit BIT	−100%	−8%	18%	31%	39%
Earnings/Equity	−5%	4%	11%	18%	25%
Retained/Equity	−10%	−2%	6%	13%	20%
Range			30%		

Table 7.5   Performance of Company with Gearing of 400 per cent (Equity =2000; Debt = 8000)

Profit BIT	500	1000	1500	2000	2500
Interest (15%)	(1200)	(1200)	(1200)	(1200)	(1200)
Profit BT	(700)	200	300	800	1300
Tax (30%)	0	(75)	(90)	(240)	(390)
Profit AT (Earnings)	(700)	(200)	210	560	910
Dividend (5%)	(100)	(100)	(100)	(100)	(100)
Retained Profit	(800)	(300)	110	460	810
Retained/Profit BIT	−160%	−30%	7%	23%	32%
Earnings/Equity	−35%	−10%	11%	28%	46%
Retained/Equity	−40%	−15%	6%	23%	41%
Range			81%		

of returns is doubled, for the assumptions adopted. The return upon equity is somewhat higher in good trading conditions, the maximum determined here being 25 per cent. Although the dividend has been assumed to be maintained even when it is not covered by profits, a repetition of such trading conditions would probably necessitate a reduction or omission of the dividend.

• For gearing of 400 per cent wide variations in outcome are predicted. In good times the return upon equity approaches 50 per cent, and substantial sums are retained for reinvestment. In bad times equity is severely eroded by interest pay-

ments; for the worst scenario about one-third of the shareholders' funds vanishes in a single year. Omission of the dividend does not offer much alleviation, since the dividend payment is not large in this case.

• In good trading conditions there is considerable scope for growth using retentions from trading profits rather than newly raised capital. Even the ungeared company has the potential for annual growth of some 10 per cent, while in the worst of the cirumstances considered it remains virtually static.

Evidently, the optimum business strategy is

to operate ungeared in bad times and with significant gearing in buoyant conditions. That is easily said, but few companies seem able to predict the onset of difficult trading conditions in time to adjust their capital structures.

8

The Grim Reaper: Taxation

Tax is indeed grim, not only in the paying but in the studying. In 1789 Benjamin Franklin observed that:

"In this world nothing can be said to be certain, except death and taxes."

He was not wholly correct, for there are many uncertainties in the taxation of corporate profits. We can agree, however, that people are certain to avoid taxes whenever possible.

The tax affairs of a multinational enterprise differ from those of private individuals in a number of important ways, all serving to render its accounts more difficult to interpret. Some features of company taxation may surprise a person whose only acquaintance with tax is his own Pay-As-You-Earn (PAYE) deduction:

- Tax is assessed not on the profit determined in the company's profit and loss statement, but on a distinct *Taxable Income* from which costs have been deducted using special procedures laid down by or acceptable to the tax authorities.
- The assessment of tax due occurs after the trading period and may be subject to subsequent revision.

- The times of payment may extend far beyond the period to which the accounts relate.
- The assessments are subject to haggling between the company's tax experts and the tax authorities, and reference may be made to tribunals or the courts to sort out unusual problems.
- Tax regulations are complex, and vary from time to time and from country to country.
- All profits of UK-based companies, wherever earned, are subject to UK tax, although double-taxation relief is usually available.

On top of all this, companies seek to arrange their affairs, at both national and international levels, in such a fashion that their overall tax bill is minimised. These financial contortions, and the currency transactions that accompany them, further obscure the tax position of a group. In consolidated accounts all of these complexities are superimposed and reported in a few lines.

Having had sight of these terrors, the reader will be relieved to learn that we shall consider only a few aspects of the taxation of an international group of companies.

8.1 Corporation Tax

We start with something fairly easy. The tax that a company pays, called *Corporation Tax* in the United Kingdom, is normally payable some nine months after the end of the tax year in which the company's accounting period ends. Hence much of the tax that is entered in the profit and loss statement for one year is actually payable in the following year. This is a simple example of what accountants call a *Timing Difference*.

Advance and Mainstream Corporation Tax

Part of the corporation tax is paid more quickly, the *Advance Corporation Tax*, abbreviated to ACT. If a company pays dividends or makes any other profit distribution to its shareholders, it must remit to the tax authorities the equivalent of basic-rate tax on the *Gross Dividend*. The 'basic rate' is that for personal UK income tax, even if the shareholder is a corporation or is resident in another country. Thus the shareholder receives the *Net Dividend*, and this investment income has in effect been brought within the PAYE system.

The entry labelled Dividends in the profit and loss statement represents net dividends. The corresponding ACT is buried within the Taxation item that appears higher up the statement. For a basic income-tax rate of 25 per cent (applied to the gross dividend), the ACT is just one-third of the net dividend. With his dividend cheque the shareholder receives a *Tax Credit* which is allowable against any UK income tax that he has to pay.

ACT is payable quarterly. But note that dividends declared in one financial year are usually not paid until the following year, and that ACT is not paid until after that. Hence even ACT carries with it a modest timing difference.

The corporation tax remaining when the ACT is deducted is termed *Mainstream Tax*; this is the element that is payable some nine months in arrears.

Overseas Taxation

As has been mentioned already, relief is usually available on tax paid abroad, although that relief is limited to the tax actually charged in the United Kingdom. Thus a company unfortunate enough to be more heavily taxed in some other country can obtain relief only up to the UK level of tax.

Nor can relief be obtained against Advance Corporation Tax. This affects companies whose income is mostly overseas, but whose dividends and ACT are paid in the United Kingdom. If this situation arises only temporarily, no great harm is done, for the relief can be moved from year to year, within limits. But a company that regularly earns much of its profits abroad may not be able to avoid paying tax twice on part of its income.

By arranging its affairs so that costs and profits arise in particular parts of the globe, a company can have some control over its tax bill. In 1991 Pilkington plc, the world's largest manufacturer of glass, whose corporate headquarters are in Lancashire, announced that its European headquarters were to be relocated in Brussels. This move is expected to reduce Pilkington's liability for unrelieved Advance Corporation Tax. However, the introduction of a unified European tax structure, which is likely before many years pass, seems likely to eliminate the advantages of such relocations between countries of the European Community.

A consequence of the interaction between ACT and overseas tax is that the earnings of a company depend on the level of the dividend that it chooses to pay. The greatest earnings arise when no dividend is paid, *Nil Basis Earnings*, and the smallest earnings when these are absorbed completely by dividend payments, *Maximum Basis Earnings*. The earnings shown on the profit and loss statement can be described more precisely as *Net Basis Earnings*. The example that follows illustrates these distinctions.

Although the possibility that the corporation tax is not large enough to offset ACT has been introduced with reference to overseas earnings, we shall encounter other circumstances in which ACT cannot be wholly set off against other tax.

Tax Losses

There are a number of ways in which a trading loss can be used to mitigate taxation. If a company records negative taxable income, the loss can be set against a profit made in the preceding year. This relief can also be used to reduce taxation on future trading profits, provided that they come from the same line of business. Relief for a loss can be transferred among the subsidiaries of a group, provided that those involved are more than 75 per cent owned by group companies. The tax losses of an acquired company move with it, to the benefit of the acquiring group, which will presumably be in profit. It will be evident that such reliefs introduce plentiful complications into company accounts.

Example 8.1 Illustrating the Effects of Tax Regime and Dividend Level

---

The hypothetical cases presented in Table 8.1 illustrate the implications for corporation tax of either declaring or not declaring a dividend, and the related implications of prior taxation of overseas earnings. The assumed tax rates are: overseas tax at 30 per cent; UK corporation tax at 35 per cent; standard rate income tax at 25 per cent. When a dividend is declared, it is taken to be 45 per cent of the Profit before Tax.

For the situation considered in the second pair of columns, two contributions to the UK corporation tax are distinguished. The third and fourth pairs of columns show the tax payable in different jurisdictions. The final columns describe a case in which insufficient relief is available to claw back taxes already assessed. In such cases the Nil Basis and Net Basis Earnings differ, as in this example.

For the situation envisaged, the Maximum Basis Earnings are 525, corresponding to total tax of 475 and a dividend of 525 that absorbs all of the earnings.

Table 8.1   Contributions to Total Tax Payable

	All UK Earned		All Overseas Earned	
	No Divd	With Divd	No Divd	With Divd
Profit before Tax	1000	1000	1000	1000
Overseas Tax	0	0	300	300
Dividend	0	450	0	450
Maximum UK CT	350	350	350	350
Double-tax Relief	0	0	300	300
Relieved UK CT	350	350	50	50
ACT	0	150	0	150
Actual UK Mainstm	350	200	50	50
	---	---	---	---
Total Tax	350	350	350	450
Profit after Tax	650	650	650	550
	Nil	Net	Nil	Net
	Basis	Basis	Basis	Basis

8.2 Deferred Taxation

Capital Allowances

It may seem surprising that, in the United Kingdom at least, depreciation is not an allowable expense for the purpose of calculating *Taxable Income*. Instead, companies deduct *Capital Allowances* at rates specified by the government of the day and subject to alteration as industrial and taxation policies change. This is equivalent to *Accelerated Depreciation* at rates above those indicative of decline in assets' utility.

Until 1984 capital allowances up to 100 per cent were allowed immediately on expenditure on plant, machinery and scientific research. The after-effects of that regime are still with us, since depreciation may extend over many years.

Currently, 100 per cent allowances are available on only a few items of expenditure, such as scientific research and buildings in enterprise zones. In addition, *Writing-down Allowances* may be applied at 25 per cent to expenditure on plant and patents, and at 4 per cent on agricultural and industrial buildings, including hotels.

Timing Differences

As writing-off capital expenditure for tax purposes commonly exceeds the rate of depreciation shown in the published accounts, capital allowances have the effect of reducing a company's taxation in the period just after it has made an allowable investment. When the allowances have been used up, the tax bill will increase again, unless further such expenditure is undertaken. Thus a portion of the tax shown in the annual accounts is *Deferred Taxation*, which may have to be paid when depreciation exceeds capital allowances. This is a timing difference of an extreme kind, whose effects may extend over many years. When the company finally pays the deferred tax, inflation will have reduced its real value. Moreover, if the money has been spent wisely, the capital asset that was acquired will be earning its keep and contributing to the cash flow needed to pay the deferred tax.

Which set of accounts is 'correct', those presented to the public, or those prepared for tax purposes? The apparent chicanery of maintaining two sets of books is explained if we remember that the company accounts are intended (we trust) to provide a realistic picture of company progress, with the smoothing provided by depreciation calculated in an invariant manner. The government does not use corporation tax merely to collect money to fund its various activities, but as a regulator of the economy and as a means of implementing industrial and social policies. By allowing accelerated writing-down of the cost of manufacturing plant and of buildings in development areas, the government provides targeted support for particular aspects of the national economy.

Provisions

How does the balance sheet respond when the actual tax payable on a year's results differs from that given in the profit and loss statement for the year? The reserves (normally that particular reserve called the profit and loss account) are depressed by the stated tax (ST), but the current assets are reduced by a smaller amount, the difference between stated tax and deferred tax ($ST - DT$). Thus the liabilities side of the balance sheet runs the risk of being too small, by the amount DT. This is corrected by introducing a provision under the heading *Provisions for liabilities and charges* equal to DT, the year's deferred tax.

The quantity DT is usually not the full amount of deferred tax, since a company's accountants may conclude that not all of it will need to be paid; at least within the three-year horizon conventionally adopted as that within which predictions can be made with some degree of confidence. If it appears likely that a part of the accumulated deferred tax will actually have to be paid during the coming financial year, it is removed from the heading of long-term liabilities and moved into the category of current liabilities.

Changes in tax law may intervene to alter the accountants' judgements, and a change in the anticipated rate of corporation tax will also require an alteration to the provision for deferred tax. Adjustments must then be entered in the profit and loss statement, to feed through to the reserves and so maintain the necessary balance of assets and liabilities.

Example 8.2 Illustrating Deferred Taxes

To demonstrate the key features of deferred tax, this artificial example considers depreciation on a straight-line basis over three years, and writing-down for tax purposes over two years, with all the deferred tax being paid in the third year. The writing-off follows the purchase of allowable equipment for £400,000. A scrap value of £40,000 is assumed; the amount to be depreciated and written-off is then £360,000.

Tables 8.2 and 8.3 indicate, respectively, the general nature of the published accounts and those kept for tax purposes, while Table 8.4 reconciles the two sets of results. Tax is reckoned at 35 per cent.

Table 8.2 Reported Profit and Loss Statements

	Year 1	Year 2	Year 3	Totals
Sales –	1000	1000	1000	3000
Depreciation	(120)	(120)	(120)	(360)
Other Costs	(600)	(600)	(600)	(1800)
Profit BT	280	280	280	840
Notional Tax	(98)	(98)	(98)	(294)
Profit AT	182	182	182	546
Depreciation	120	120	120	360
Cash Flow	302	302	302	906

Table 8.3 Actual Write-offs and 'Cash Flows'

	Year 1	Year 2	Year 3	Totals
Sales	1000	1000	1000	3000
Write-off	(180)	(180)	–	(360)
Other Costs	(600)	(600)	(600)	(1800)
Taxable Income	220	220	400	840
Actual Tax	(77)	(77)	(140)	(294)
Profit AT	182	182	182	546
Write-off	180	180	–	360
Cash Flow	323	323	260	906

Table 8.4 Reconciliation showing Deferred Tax

	Year 1	Year 2	Year 3	Totals
Reported Tax	(98)	(98)	(98)	(294)
Deferred Tax	21	21	(42)	–
Actual Tax	(77)	(77)	(140)	(294)
Tax Provision	21	42	0	–

Table 8.3 shows an improved 'Cash Flow' for the years just after the investment is made, as a result of the excess of capital allowances over depreciation. The extra funds might be used to pay interest on the borrowings used to buy the equipment, or to begin to pay off the debt. Table 8.4 shows how the Provision is built up and then discharged in Year 3.

Example 8.3 Southern Water plc 1991: Capital Allowances and ACT

The water supply and sewerage companies were floated on the stock market in 1989, following the decision of their former owner, HM Government, that they should be privatised. The companies were allowed to carry forward capital allowances in respect of earlier expenditure on plant, and were granted such allowances on the very large expenditures known to be necessary to meet the higher standards of water purity now imposed. The consequence of these reliefs is that these companies will not have to pay mainstream corporation tax for many years, probably not at any time in the twentieth century.

We here consider one of these companies, Southern Water, which operates in Kent, Hampshire, Sussex and the Isle of Wight. Its tax affairs are particularly simple, as it has no overseas earnings and, just privatised, is starting off with a clean slate. For the year 1991 it declared dividends of £29.0 million and taxation of £9.7 million, and commented on its tax position as follows:

"The Group has available £15.2 million (1990: £5.5 million) of unrelieved advance corporation tax and substantial unutilised capital allowances. At 31 March 1991 unutilised allowances existed in respect of expenditure qualifying for plant and machinery allowances of approximately £522 million and £450 million in respect of expenditure qualifying for industrial building allowances.

"Until such time as a foreseeable liability to mainstream corporation tax arises, the only tax charge to the profit and loss account will be in respect of advance corporation tax."

This statement illustrates a number of the points made in general terms earlier. Note that the tax due is just one-third of the dividends paid, as it consists only of Advance Corporation Tax.

Example 8.4 British Steel plc 1988/89:
Tax Losses and Deferred Taxation

For the financial year ending 1 April 1989 British Steel reported Profit on ordinary activities before taxation of £593 million and Tax on profit on ordinary activities of £31 million. At only 5 per cent of profit this tax is far below the 35 per cent rate of corporation tax that pertained in that year. British Steel explains how this came about in Notes to the accounts.

Accounting Policy on Taxation:

"The charge for taxation is based on the profit for the period as adjusted for disallowable items. Tax deferred or accelerated is accounted for in respect of all material timing differences to the extent that it is probable that a liability or asset will crystallise. Timing differences arise from the inclusion of items of income and expenditure in tax computations in periods different from those in which they are included in the accounts. Provision is made at the rate which is expected to apply when the liability of asset crystallises."

Explanations of Tax Calculations:

"(i) The tax charge for the year is reduced by the utilisation of tax losses of £81 million brought forward and the use of previously deferred capital allowances of £453 million. These two items had the combined effect of reducing the tax charge for the year by £187 million.

"(ii) ACT written off as irrecoverable in 1974 has now been written back as available to set against taxation on current profits.

"(iii) Accumulated tax losses available at 1 April 1989, after taking into account a reduction of £1699 million under Section 11(3) of the British Steel Act 1988 and claiming in full capital allowances deferred in earlier years, are estimated at £90 million.

"(iv) The full potential amount of deferred taxation calculated at 35% on all timing differences is . . . £139 million. No deferred tax has been provided as projections indicate that the potential liability will not crystallise in the foreseeable future."

Adding back (to the declared tax of £31 million) the £187 million reduction in tax charge determined in Note (i), we return to the level of 35 per cent that might be expected. As anticipated in Note (iii), British Steel did exhaust its tax losses in the following year, 1989/90; thereafter this form of relief will not be available to it. The reduction in reliefs mentioned in Note (iii) compensates for the government's investment of about that amount in British Steel a few years prior to privatisation. Finally, in the year 1989/90 British Steel did find it appropriate to make a provision for deferred tax, in respect of prepayments of pension contributions.

Example 8.5 Imperial Chemical Industries PLC 1990:
Presentation of Tax Affairs

This is often described as Britain's largest manufacturing company, although it should be noted that of the year's Turnover of £15,379 million (before inter-area eliminations) only £6126 million was generated by UK subsidiaries, and of that their overseas sales amounted to £3160 million.

In its best years ICI has declared Profit BT around £1500 million, and Profit AT approaching £1000 million. The year considered here was not one of the best; Profit BT was £977 million and the reported tax charge was £338 million. This happens to be about 35 per cent of the declared profit, but an examination of the Notes accompanying the accounts shows that this is the net result of several compensating contributions.

ICI makes a considerable effort to explain its tax affairs in the Notes set out below. They are reproduced here to cast light on aspects of company taxation in general. Some of the notes give information on the tax affairs of the parent company as well as the group. We shall consider only the group values.

Note 5. *Tax on Profit on Ordinary Activities:*
ICI and subsidiary undertakings

United Kingdom taxation:	
Corporation tax	109
Double Taxation relief	(43)
Deferred taxation	2
Overseas taxation:	
Overseas taxes	186
Deferred taxation	39
	___
	293
Associated undertakings	45
	___
Tax on Profit on ordinary activities	338

UK and overseas taxation has been provided on the profits earned for the periods covered by the Group accounts. UK corpora-

PLATE 11

ICI's KLEA 134a plant at Runcorn, Cheshire. KLEA is the first commercial CFC alternative.

tion tax has been provided at the rate of 35 per cent.

As would be expected, Note 5 shows that overseas tax is very important for ICI, which does not appear to have obtained relief on much of that tax. Some previously

Note 5. *Deferred taxation*		
Accounted for at balance sheet date	1990	1989
Timing differences on UK capital allowances and depreciation	63	53
Miscellaneous timing differences	10	125
Advance corporation tax recoverable	(80)	(79)
	___	___
	(7)	99
Not accounted for at balance sheet date		
UK capital allowances utilised in excess of depreciation charges	330	324
Miscellaneous timing differences	(65)	16
	___	___
	265	340

deferred tax is now being paid, both in the UK and more especially abroad. An equity-determined proportion of the tax of Associates appears, since that proportion of their profits has been entered in the group's profit and loss statement.

The second extract from Note 5 includes values for the preceding year, to illustrate the significant changes that can occur from one year to the next. An unusual feature of the 1990 'provision' for deferred taxation is that it is negative. The anticipated benefits of recovering ACT are larger than what are believed to be the potential liabilities. It will be apparent that ICI expects to pay only a small fraction of the total tax that has been deferred. Possibly this is merely a consequence of the three-year horizon adopted in defining this kind of liability.

To put deferred taxation into context, we look in Note 18 at all the contributions to the Provisions. Other notes (not presented

Note 18. *Provisions for liabilities and charges*

Deferred taxation	(7)
Employee benefits	212
Reshaping, environmental and other provisions	344
Total	549

here) give information on unfunded pension costs that contribute to the Employee Benefits item, and on the extraordinary items that influence Deferred Taxation. The latter include a number of 'restructuring' activities intended to reduce ICI's cost base.

It will be apparent that someone with sufficient understanding of methods of accounting for tax can gain considerable insight into ICI's tax position, using the substantial body of data contained in this series of cross-referenced notes. Not all companies provide as much information on this aspect of their finances.

9

Analysis of Accounts

This may seem an odd title, since many aspects of company accounts have been dissected in the examples of the preceding chapters. Here we adopt a more systematic approach, and introduce the measures of performance in general use for financial analysis. It is unrealistic to expect to find a few parameters that will adequately describe the workings of a group of varied businesses influenced by a stream of national and international events. Indeed, the need to introduce a large number of indicators reveals the absence of consensus on the parameters of greatest significance.

Some of the questions for which we would like answers are:

- How does the group compare with others in its line(s) of business, that is to say, with its competitors?
- How does it compare with companies engaged in quite different activities?
- Is the company making money as rapidly as might be expected?
- Will it be able to maintain the payments on its borrowings?
- Will the company be able to pay its other bills as they come due?
- Are its operations efficient, by a variety of criteria?
- Does the company employ its staff effectively and support them adequately?
- What does the group's future look like?
  And finally:
- Should I invest my money here?

The varied performance indicators defined in this chapter, mostly in terms of numbers provided in company reports, bear upon the questions posed above. Even when these parameters are available, further analysis and the exercise of judgement are needed to reach meaningful conclusions about the health of a company's finances and commercial operations.

Sir Peter Swinnerton-Dyer has commented ruefully on the use of performance indicators to monitor the affairs of universities, but his assessment has wider application:

"A performance indicator is a number which can be calculated by a good statistician without any exercise of judgement and which is a surrogate for a measurement of what one is actually interested in. Unfortunately, outsiders are apt to believe that they provide an adequate means of assessing a system without the labour of understanding it."

9.1 Ratio Analysis

Comparisons of company performance must allow for differences in the scale of their activities. Moreover, looking beyond the United Kingdom, we encounter financial data expressed in varied currencies, influenced by distinctive taxation regimes, and drawn up using different accounting conventions. The problems of size and currency are addressed by looking at scaled values, that is, ratios of quantities that have the same units, commonly of the form Money/Money.

Common-Size Analysis.   A useful step is division of every number in a column of the accounts by the biggest number, so that the others fall into place as fractions or percentages. For example, all the entries of the profit and loss statement can be divided by turnover, the largest number there. Dividing all the entries of the balance sheet by capital employed, we achieve a similar standardising effect. These processes remove the major effects of scale and allow organisations of very different size to be compared.

Standardised Ratio Generation.   The process can be carried further, by creating a cascade of ratios indicating, for example, the progressive changes that take place on descending through the profit and loss statement. A Centre for Interfirm Comparisons regularly generates standardised values in this way. It can produce the complete array of ratios, only for companies that co-operate, by providing details in a specified form and beyond those normally given in published accounts. Of course, this process may generate so many numbers that the significant features remain hidden in the haystack. Moreover, meaningful comparisons are difficult for companies with multiple products, complex webs of joint ventures, and differing geographical ranges.

Some commercial organisations regularly assess their operations against a series of performance indicators. The now-disbanded Central Electricity Generating Board, for example, sought to control costs using an array of forty-one performance indicators. Fig. 9.1 shows a corner of the array. Note that some of the ratios are not dimensionless, but are produced by dividing costs by measures of physical output (in this case, kW h of electricity sold) or measures of production capacity (in this case, the attainable output of CEGB plants in MW).

Time-Series Analysis.   Companies report what they consider to be key financial quantities relating to each of the last five years, and sometimes longer. Perhaps there is little point in looking further back, for the business environment is seldom stable for even five years, and the composition of a group may change so much that the entity considered at the beginning of the period is hardly recognisable at the end. In the five years 1986 to 1990 British Aerospace, for example, absorbed Royal Ordnance, a Dutch construction company, the Rover Group, a property-development company, and numerous smaller fry, thus increasing its turnover by some 230 per cent.

Nevertheless, one can learn from the past. A single year's poor performance might be massaged away by adjusting provisions, or by bringing forward income that has been left undeclared for just this purpose. A succession of bad years is more difficult to hide.

Cross-Sectional Analysis.   By taking a section through an entire industrial sector, one can determine industrial norms and can identify the organisations that depart significantly from them. In some sectors appropriate comparators are few. ICI, for instance, is itself a significant fraction of the United Kingdom's chemical industry. BAA plc – it owns the main London airports, very much the busiest in the country – has no obvious comparator, anywhere in the world.

Even when appropriate comparators exist, information about them may be hard to find. Consider the British motor industry. Rover is buried within British Aerospace, Rolls-Royce within Vickers, Vauxhall within General Motors, Peugeot-Talbot

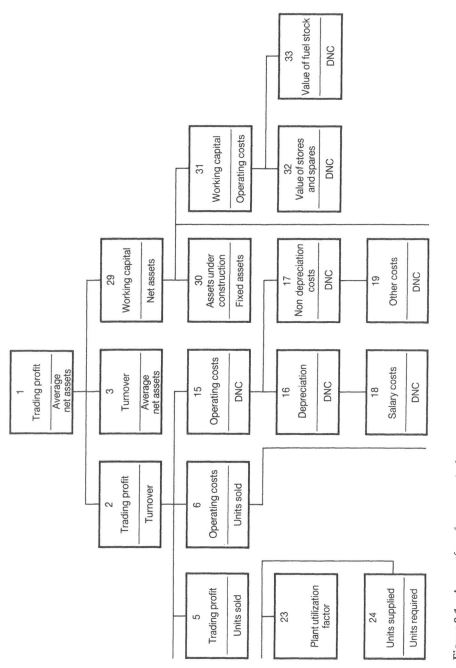

Figure 9.1 Array of performance indicators adopted by the Central Electricity Generating Board. 'Units' means output in kWh. 'DNC' is declared net capability, measured as MW sent out

within Peugeot SA, Ford UK within the international Ford group, and Jaguar within Ford UK. Conglomerates are now required to report more explicitly the affairs of the distinct businesses buried within their accounts, and we can hope to receive increasingly more complete information as the years pass.

In a rapidly changing commercial environment differences in reporting dates can degrade comparisons to an important degree. This difficulty was expressed by Sir Antony Pilkington in a letter to *The Economist*, complaining that its comparison of the performance of a number of glass-making companies was unfair because the end of his company's reporting year was March 1991, while the accounting periods adopted by the four comparators ended in December 1990 and March 1990. He pointed out that the Gulf war and "gathering recession in western economies" profoundly altered the performance of the other companies in the periods just after those considered in the article.

Example 9.1 Ford Motor Company, AB Volvo, Jaguar plc 1988:
Scaled Comparisons

Comparisons of the performance of these automotive companies are complicated by a number of factors. The companies are of very different size, and they report respectively in American dollars, Swedish crowns, and pounds sterling. We note also that, although Jaguar is a 'pure' car company, both Ford and Volvo engage in other businesses.

Table 9.1 presents the results reported by the three car makers. The year 1988 has been chosen as it was the last in which Jaguar was independent and reported publicly. The figures given for Ford exclude its significant Financial Services operations, but do include such activities as Ford Aerospace and Rouge Steel. The Volvo results are those for the entire group and include contributions from food processing, aerospace products, and so forth.

Table 9.2 presents the same information in scaled fashion, with Sales as the scaling factor. Here differences in performance stand out quite clearly. It must be remembered, however, that accounting conventions and tax regulations vary from country to country. Great care is required to construct well-justified comparisons.

Table 9.1   Basic Income Statements

	Ford Motor (US$ mill)	AB Volvo (SwK mill)	Jaguar (£ mill)
Sales	82193	96639	1076
Costs	75581	89431	937
Operating Income	6612	7208	139
Profit BT	7312	5932	48
Net Income	4609	3329	28

Table 9.2   Scaled Income Statements

	Ford Motor	AB Volvo	Jaguar
Sales	100	100	100
Costs	92.0	92.5	87.1
Operating Income	8.0	7.5	12.9
Profit BT	8.9	6.1	4.5
Net Income	5.6	3.3	2.6

9.2 Profitability: the Bottom Line?

Margins and Returns.   Some often-asked questions concerning profitability are: 'How good is the company at converting sales volume into profit?' 'How well does it use its assets?' 'How well does it use the funds invested in it?' Ratios that provide insight into these matters are:

Profit Margin on Sales = Profit/Sales

Return on Assets = Profit/Total Assets

Return on Equity = Profit/Shareholders' Funds

Return on Capital Employed (ROCE)
   or Return on Investment (ROI) = Profit/Capital Employed

Note that ratios relating profits to sales are called *Margins*, while those relating profits to capital are called *Returns*. The latter were introduced earlier through Equations (7.8) and (7.9).

We can relate margins and returns using a little algebra:

$$\text{Profit/Assets} = (\text{Sales/Assets}) \times (\text{Profit/Sales})$$

$$\text{or Return on Assets} = \text{Asset Utilisation} \times \text{Profit Margin}$$

Any one of the measures of assets (equity, total assets, capital employed = net assets) can be used in this decomposition. If profits fall, this simple analysis distinguishes between possible causes, either less effective use of capital to generate sales, or failure to convert turnover into profit.

To assess the performance of a company relative to others in its industry, we can prepare a plot of the form shown in Fig. 9.2. The straight line represents combinations of factors that give the same return, a return achieved by the better-performing companies of the class considered. Companies whose current operations are represented by points below that line will presumably seek to move towards it. Seemingly, the strategy appropriate to the circumstances of an individual company will depend on

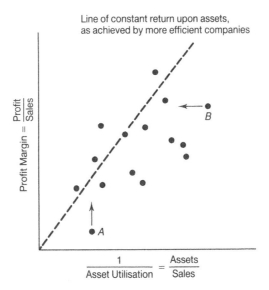

Figure 9.2   Analysis of return upon assets. The points represent the performance of companies in the same line of business.

its position relative to the more successful companies in its class. The arrows suggest plausible aspirations for laggards *A* and *B*.

Measures of Value Added. It can be argued that our attention is better focused on a company's ability to create added value and to convert it into profits.

These aspects can be assessed by considering:

$$\text{Value-Added Margin} = \text{Value Added/Sales}$$

$$\text{Margin on Value Added} = \text{Profit/Value Added}$$

These are linked to returns and profit margin by

$$\text{Profit/Assets} = (\text{Value-Added/Assets}) \times (\text{Profit/Value Added})$$

$$\text{Profit/Sales} = (\text{Value-Added/Sales}) \times (\text{Profit/Value Added})$$

The Value-Added Margin is a measure of the degree of *Vertical Integration* of a group's activities. The ratio will be lower for an organisation that merely distributes than for one that manufactures components, assembles them into completed products, and even controls the retail outlets through which they are sold. The major oil companies, such as the British Petroleum Company, are classic examples of strong vertical integration.

Choice of Measure of Profit. We noted above that a variety of measures of assets can be used in defining returns. We now consider which of the measures of profit should be used in defining margins and returns. If we wish to investigate the performance of a company's production unit, it is appropriate to select some measure of profit before interest and tax. If we wish to assess the overall performance of the company, including its handling of debt and taxation, earnings will be the proper choice. Logically, in calculating Return upon Equity, we should use profit to ordinary shareholders.

Finally, we note that the ratio

Retained Profit/Shareholders' Funds

gives some indication of a company's ability to grow without seeking external funding. In Sections 9.7 and 9.8 other parameters bearing on this question will be introduced.

Example 9.2 The Electricity Industry 1989/90: Profit Margins

During the years 1987 to 1991 the business of generating and distributing electricity within the United Kingdom underwent radical restructuring. The activities that had taken place under The Electricity Council were progressively split into distinct businesses, some in competition, and many were then floated on the stock market. These processes provide an opportunity to see how the parts of this industrial sector mesh together.

The generating business of England and Wales (originally the province of the Central Electricity Generating Board) was split into three parts: National Power plc, PowerGen plc and Nuclear Electric plc. (The last was not floated on the stock market, in view of its hard-to-quantify liabilities relating to the decommissioning of nuclear power stations.) The distributing of electricity

to individual consumers was passed to twelve regional distribution companies, who jointly own the National Grid Company plc, which operates the core of the transmission network.

The results reported by these companies in the year after their formation are given in Table 9.3. Costs were apportioned to the successor companies in a somewhat arbitrary manner, but with the intention of indicating how the costs were likely to fall when the bits of the jig-saw puzzle were finally taken apart. The costs of Nuclear Electric are particularly difficult to assess realistically.

On comparing the profit margins, we should not jump to the conclusion that the companies further down the line – the transmitters and distributors – are less profitable in a fundamental sense. For a

Table 9.3   Operating Profit Margins for 1989/90 (values in £ million)

	Sales	Costs	Margin
National Power plc	3998	3542	11.4%
PowerGen plc	2608	2217	15.0
Nuclear Electric plc	2058	1833	10.9
Total for Generators	8664	7592	14.0
National Grid Company plc	9735	9305	4.4
Distributing companies	13324	12492	6.2

number of reasons they neither need nor are able to justify large profit margins. They do not have the large commitments to capital expenditure that will fall upon the generators, in building gas-fuelled plant and in reducing the environmental effects of their activities. They are in the typical position of distributors and retailers, buy-ing a product whose final value has been substantially determined, and conveying it to the consumer. While their sales are large, their costs – other than those of purchasing the commodity – are relatively small. On simple arithmetic grounds this leads to a smaller profit margin.

Example 9.3 Yorkshire Water plc 1990/91, Scottish Hydro-Electric plc 1990/91, J Sainsbury plc 1991: Value-Added Ratios

---

Each of these companies is in the water business. Yorkshire Water collects, treats and distributes water in its part of Eng-land, and then collects, treats and disposes of waste water. Scottish Hydro-Electric (which prefers to be known simply as Hydro-Electric and is the successor of the North of Scotland Hydro-Electric Board) generates, transmits and retails electricity, though it should be noted that less than 20 per cent of its generating capacity is actually water-driven. The greater part of many of the commodities sold in Sainsbury stores is actually water.

Table 9.4 gives key figures from the profit and loss statements of these companies. The first reckoning of Cost of Sales is that leading to Gross Profit, and the second gives Operat-ing Profit. Yorkshire Water does not provide the information required to determine gross profit.

Table 9.5 presents derived performance-indicating ratios. Consideration of the three margins on sales calculated for Scottish Hydro-Electric shows how important it is to know which measure of profit is being considered. Sainsbury's margins on sales are much lower than those of the other com-panies, since its operations cover a smaller part of the production/distribution chain. In fact, among food retailers Sainsbury is renowned for squeezing healthy profits from its sales.

The Value-Added Margins are ranked as one would expect, in view of the varying degrees of vertical integration of the com-panies' operations. In a sense, Yorkshire Water's raw material is free, coming to it either from the clouds or down the drain. Much of 'Hydro-Electric's' elec-tricity comes from the burning of fossil fuels; thus the cost of its raw materials is

Table 9.4 Sales, Profits and Values Added (in £ million)

	Yorkshire Water	Scottish Hydro-Electric	Sainsbury
Sales	389	566	7813
Cost of Sales (1)	–	326	7050
Cost of Sales (2)	283	449	7229
Value Added	237	231	1590
Gross Profit	–	240	763
Operating Profit	106	119	585
Profit BT	114	60	518
Earnings	103	49	355

Table 9.5 Performance Indicators

	Yorkshire Water	Scottish Hydro-Electric	Sainsbury
Gross Profit Margin	–	42.4%	9.8%
Operating Profit Margin	27.2%	21.0	7.4
Earnings/Sales	26.5	8.7	4.5
Value-Added Margin	60.9%	40.8%	20.4%
Operating Profit/Value Added	44.7	51.5	36.8
Earnings/Value Added	43.5	21.2	22.3

a larger fraction of turnover. In neither case is a distribution and retail chain interposed between them and the consumer.

When we look at Margin on Value Added (Operating Profit/Value Added) the performance of the three companies is more nearly similar. Indeed, the capital-intensive Scottish Hydro-Electric looks best by this criterion. One reason for Sainsbury's lagging behind the other two may be their monopolistic or semi-monopolistic positions in their markets.

PLATE 12

Sloy hydro-electric power station on Loch Lomondside.

9.3 Liquidity: Can the Bills be Paid?

Here we consider current assets, particularly cash and investments readily converted into cash, that may be available to pay off a company's short-term liabilities. In reality, when a company ceases to trade, its remaining assets are usually absorbed by long-term creditors, who have a priority claim. Hence it is more realistic to interpret measures of liquidity as indicators of the ability of a company to pay its bills promptly and the freedom of its managers from nagging concerns about the cash position.

Although balance-sheet values are not always good indicators of the availability of cash and likely demands for it, they are usually all we have to work with. Two ratios are commonly considered:

Current Ratio or Working Capital Ratio

= Current Assets/Current Liabilities

Liquid Ratio or Quick Ratio or Acid Test

$$= \frac{\text{Current Assets - Stocks}}{\text{Current Liabilities}}$$

The latter recognises that Stocks (valued on a going-concern basis) may not be convertible into cash in an emergency. It is important to check the precise definitions used in determining both assets and liabilities; sometimes short-term loans are omitted from the liabilities.

The conventional wisdom is that Current Ratio = 1.5 to 2 is normal. A value above 3 may be thought to suggest over-liquidity, that is, too much money tied up in stocks, too many accounts outstanding, or too much cash merely earning interest. In other words, the working capital is too large.

For the Acid Test, the conventional wisdom is that Liquid Ratio = 1 is about right. A value below 0.7 is sometimes thought to be a sign of danger.

The normal values of these ratios are strongly dependent on the fundamental nature of the business in which the company is engaged. Hence, changes are probably of greater significance than absolute values. Even the average values for industrial sectors move significantly as one passes through the trade cycle. In 1983 the liquid ratio for manufacturing companies in the United Kingdom was around 1.1. It fell below 0.7 in 1985, rose above 0.9 in 1987, and thereafter fell to below 0.5 as recessionary conditions developed.

Another way of assessing the level of working capital is through

Working Capital Ratio = (Net Current Assets)/SHF

This indicates the fraction of the equity that is tied up in working capital. The use of this ratio suggests that borrowings are perceived to fund capital expenditure, while shareholders provide the funds to keep the business running.

Example 9.4 British Aerospace plc 1990, Glaxo Holdings plc 1990,
J Sainsbury plc 1991: Measures of Liquidity

In Example 5.4 we considered the current assets and liabilities of these companies, identifying the quantities that contribute to the conventional measures of liquidity. Table 9.6 presents the ratios obtained using those numbers and the SHF for the three companies of £2534, £2732 and £1672 million, respectively.

The most important lesson to be drawn from these comparisons is that even stable companies display very different values of the liquidity ratios. In particular, Sainsbury operates with negative working capital (as was noted in Example 5.4), and this leads

Table 9.6 Liquidity Ratios

	BAe	Glaxo	Sainsbury
Current Ratio	1.30	2.05	0.43
Liquid Ratio	0.69	1.76	0.18
Working-Capital Ratio	0.55	0.51	−0.49

to very low values of the current and liquid ratios, and obviously to a negative value of the working-capital ratio.

9.4 Efficiency: Speed and Control

Turnover Ratios. These are indicators of the speed with which a company performs key operations within its operating cycle and of its control over working capital. It may not be helpful to suggest typical values, since these measures are strongly dependent on the nature of the commercial activity which they describe. Moreover, parameters calculated using gross values representing a diverse group are of doubtful significance. Whenever possible, values characterising a particular sector or product should be used.

The number of times that stocks were replenished during the year is given by

Stock Turnover Ratio = (Cost of Sales)/Average Stocks

This measures a company's control over its inventory. A high value, for the particular kind of business, is desirable; it implies that materials do not sit around before being turned into saleable products, and that completed products quickly go off to fill orders. For manufacturing industry a value or 3 or 4 has been common, but the introduction of *Just-in-Time Stock Control* can be expected to produce higher values.

The ability to generate sales from capital investment is measured by

Fixed-Assets Turnover Ratio = Sales/Fixed Assets

Again, a high value is desired. We noted earlier (Section 9.2) that this factor contributes to returns on assets.

Since 'working capital' is tied up in the business, rather than invested in productive plant, one seeks to minimise it and to convert it into sales as often as possible. A company's effectiveness in doing this is measured by

Turnover of Working Capital = Sales/Working Capital

However, a very low value of working capital (= current assets − current liabilities) would conflict with the conventional wisdom on Liquidity.

Looking at the components of working capital, we can define other instructive quantities, such as:

Debtor Turnover = Sales/Average Trade Debtors

This is a measure of the control a company exercises over its debtors, that is, the speed with which it extracts payments from those who owe it money. On the other hand

Creditor Turnover = Sales/Average Trade Creditors

is the rate at which a company's suppliers extract money from it. In hard times this quantity may be allowed to increase, thus transferring the pain to suppliers. In periods of recession this process creates serious difficulties for small businesses.

Turnover Periods. Any of the turnover ratios can be converted into the period required to carry out the relevant activity. Thus we have

Credit or Collection Period = 365 /Debtor Turnover   (days)

The other ratios can be treated in a like manner, to determine the average number of days that a company's bills go unpaid, or that its stocks remain unprocessed or unsold.

A measure of the length of a company's *Operating Cycle* is the sum

Stock Turnover Period + Debtor Period

This indicates in a general way the time between receipt of materials and receipt of payment for products made from those materials.

Overhead Control. One is naturally interested in the degree to which a company's resources are devoted to activities extraneous to the central function of meeting the needs of its customers. We can define

An Overhead Ratio = (Element of Costs)/Total Costs

The Element of Costs might be the expenses of Administration, Distribution, or Research and Development. Of course, engineers are likely to argue that R&D is not an 'overhead' but the future of the business.

A somewhat similar measure of financial efficiency is

Tax Ratio = Tax/Profit BT

Some international groups work wonders in reducing their overall tax commitments. But a modest tax bill is not always a good sign. As we saw when considering British Steel in Example 8.4, low tax outgoings may be the consequence of loss-making in earlier years.

An indicator of a company's success in managing its debt, and of its standing in the investing community, is the

Average Interest Rate = Total Interest/(Average Total Debt)

A company that is not under financial strain is able to keep this average low, by choosing times of relatively low interest rates to issue long-term loan securities.

Example 9.5 British Aerospace plc 1990, Glaxo Holdings plc 1990,
J Sainsbury plc 1991: Stock Turnover

We turn to these companies to illustrate yet another aspect of company finance. In Table 9.7 the stock turnover ratio and the related period are derived from the relevant accounting data. The net Cost of Sales given here has been found by deleting some income that is extraneous to basic trading activities.

By the conventional criteria for manufacturing concerns, British Aerospace and Glaxo Holdings appear to be doing quite well. As would be expected from the nature

Table 9.7 Stock Turnover and Stock Turnover Period

	BAe	Glaxo	Sainsbury
Average Stocks	2596	372	335
Cost of Sales	10420	1902	7050
Turnover Ratio	4.0 p a	5.1 p a	21.0 p a
Turnover Period	90 days	70 days	17 days

of its business, Sainsbury turns around its stock much more quickly.

9.5 Productivity: Is the Labourer Worthy of his Hire?

Most of the ratios introduced above are of the form Money/Money. Productivity measures are often 'per employee' quantities. Some that are useful in making comparisons within a particular industrial sector are:

Turnover per employee
Value Added per employee

Employee costs per employee (that is, average salary plus other benefits)

Because they are dependent on regional and social factors, these ratios can be misleading if applied, without due thought, to an international range of activities. Boards of directors consider them very carefully when deciding how to distribute their manufacturing around the world.

Two dimensionless measures of productivity are

$$\text{Sales/(Costs of Employment)}$$

$$\text{Value Added/(Costs of Employment)}$$

There is a better chance that these will be applicable from country to country. The second ratio can be split into

$$\frac{\text{Total Assets}}{\text{Costs of Employment}} \times \frac{\text{Value Added}}{\text{Total Assets}}$$

or   Capital Intensity × Asset Utilisation

Each of these factors measures a kind of productivity.

Example 9.6 British Aerospace plc 1990, Glaxo Holdings plc 1990,
J Sainsbury plc 1991: Employee-Related Parameters

Looking again at these three companies, we assess their effectiveness in the use of employees and other assets. Table 9.8 presents the relevant information (values in the upper part are in £ million) and the ratios flowing from it. In the lower part of the table Assets* means Total Assets less Current Liabilities.

Table 9.9   Measures of Productivity and Asset Utilisation

	BAe	Glaxo	Sainsbury
Turnover	10540	2854	7813
Value Added	3123	1902	1589
Costs of Employment	2258	645	879
Employees (1000 × full-time equivalents)	127.9	31.3	70.8
Total Assets less Current Liabilities	4666	3030	2420
Sales/Employees	£82400	£91200	£110400
Value-Added/Employees	£24400	£60800	£22400
Assets*/Employee	£36300	£96800	£34200
Average Benefits	£17650	£20600	£12420
Sales/Empl. Costs	4.67	4.42	8.89
Value-Added/Empl. Costs	1.38	2.95	1.81
Assets*/Empl. Costs	2.07 yr	4.70 yr	2.75 yr
Sales/Assets*	2.26 pa	0.94 pa	3.23 pa
Value-Added/Assets*	0.67 pa	0.63 pa	0.66 pa

Most features of these comparisons are as one would expect. Sainsbury has the highest Sales/Employee and the lowest average rewards to its employees. It is perhaps a little surprising that Glaxo's staff are better rewarded than those of British Aerospace. The employees of Glaxo and Sainsbury appear to be better supported by Assets than are those of British Aerospace. The fraction of the Value Added that is absorbed by Costs of Employment varies significantly, from 72 per cent (BAe) to 34 per cent (Glaxo).

When comparing these values, we should remember that the fixed assets of British Aerospace are probably undervalued. Consistent with this, the values of Assets/Employment Costs and of Assets/Sales seem unrealistically low in comparison with those of Glaxo Holdings. If this value were brought into line with Glaxo's, the Value-Added/Assets ratio would be depressed relative to those of the other companies.

9.6 Capital Structure: Strain Gauging

In Section 7.2 we looked at the balance between debt and equity in a company's capital, and introduced the term *Gearing* to specify it. A more precise definition is

$$\text{Capital Gearing} = \frac{\text{Long-term Liabilities (including Preference Stock)}}{\text{Ordinary Shareholders' Funds}}$$

A second ratio does much the same job but is less often encountered:

$$\text{Capital Gearing} = \frac{\text{Long-term Liabilities (including Preference Stock)}}{\text{Total Assets} - \text{Intangibles}}$$

As was mentioned earlier, the term *Leverage* is used in other countries, for either of these quantities.

The definition of gearing (or leverage) needs to be watched carefully. Some companies will argue that their Intangibles are just as Fixed as their other assets. Others hold that Net Liabilities should be used in calculating gearing, that is, Liabilities less liquid assets.

The capital structure of Thorn EMI was considered in Example 7.2, and the changes in these two measures of gearing were determined for each year within a five-year period. In those calculations Thorn EMI measured its indebtedness using *Net Borrowings*, that is, long-term borrowing less the liquid component of current assets. It is debatable whether this provides greater insight into the company's financial position. Certainly it does not specify its capital structure as clearly as do the ratios defined above.

The ratio of the two measures of capital gearing is the

$$\text{Proprietary Ratio} = \frac{\text{Ordinary Shareholders' Funds}}{\text{Total Assets} - \text{Intangibles}}$$

This tells us how much of the company is actually owned by its owners. This ratio also equals

$$\frac{\text{Return upon Total Capital}}{\text{Return upon Equity}}$$

Thus it shows how gearing acts to lift the Return upon Equity above the Return upon Capital Employed.

Why is the name 'capital gearing' applied to what was earlier simply called 'gearing'? Because we may also consider

$$\text{Income Gearing} = \text{Interest Paid/Profit BT}$$

(Net Interest Expense is sometimes used instead.) This indicates how easily the company could pay last year's interest on its debts. More to the point, it suggests how far profits would have to drop before interest payments really begin to hurt.

Conventional wisdom is that

$$\text{Income Gearing} \quad < 2 \text{ suggests over-borrowing}$$
$$> 6 \text{ suggests under-borrowing}$$

It is the former condition that is likely to excite a company's bankers. Some analysts hold that income gearing is a better measure of debt level than capital gearing. They believe that it is a more direct measure of a company's ability to stay out of trouble.

In Section 7.2 the 'normal' range of capital gearing was indicated. The suggestions made there, like those just made in regard to income gearing, do not apply uniformly to all business sectors. It is unjustified changes in gearing that should be watched for.

9.7 Stability: Here Today and Gone Tomorrow?

The likelihood that a company will grow steadily, or as steadily as the economic environment permits, can be assessed using a variety of factors. Some are expressed in straightforward numerical terms; others are more subjective and require the exercise of judgement. The questions to be considered here relate also to Investment Potential, which will be discussed in Section 9.9.

Quantitative Factors. The investor will want to know the extent of the firm orders a company has on its books – in American parlance the *Backlog*. This can be expressed as a period of time:

$$365 \times (\text{Orders on Hand})/\text{Sales} \quad (\text{days})$$

One likes to see this increasing from year to year.

A knowledge of the sensitivity of profits to a reduction in sales volume is useful. We can define

$$\text{Income Sensitivity} = (\% \triangle \text{ Profits})/(\% \triangle \text{ Sales})$$

$$= \frac{\text{Profit BIT}}{\text{Profit BT}} \times \frac{\text{Sales - (Variable Costs)}}{\text{Sales - (Variable + Fixed Costs)}}$$

$$= \quad \text{Financial Leverage} \times \text{Operating Leverage}$$

The variable costs are usually taken to be directly proportional to sales volume.

The first factor tells us how the profit sensitivity is influenced by interest payments; the Financial Leverage >1, unless the company has negligible or negative borrowings. The Operating Leverage indicates the effect of a change in the level of sales, with part of the cost of sales remaining fixed. The proportional change in profits equals that in sales if both interest and fixed costs are negligible. Since these conditions are not usually met, profits normally change more rapidly than sales.

Few companies distinguish in their reports the fixed and variable components of their costs. Indeed, it is no easy matter to determine these costs within a business. If the corporate and capital structures remain constant, an estimate can be made by plotting sales against profits for a number of reporting periods. We can define

$$\text{Margin of Safety} = \text{Sales - Fixed Costs - Variable Costs} \ (\propto \text{Sales})$$

As sales fall, this margin falls. The level of sales at which it vanishes is termed the *Break-Even Point*.

The relationship between capital expenditure and depreciation indicates the extent to which a company's asset base is being maintained. When the ratio

$$\text{Capital-Spending/Depreciation} < 1$$

it would seem that the asset base is falling. On the other hand, a value exceeding unity suggests that depreciation will rise in the years to come and that profits will be depressed in consequence. In considering this relationship, we must try to allow for the effects of inflation.

A second parameter indicates whether an adequate charge against profits has been made:

$$\text{Depreciation Expense/Depreciating Assets}$$

Changes in this ratio may reveal attempts to 'smooth' the reported profits by adjusting depreciation rates.

An indicator of a business's need for external capital is the

Funds Flow Adequacy Ratio

$$= \frac{\text{Five-year sum of Funds from Operations}}{\text{Five-year sum of Capital Spend + Change in Stocks + Dividends}}$$

Obviously, this parameter allows for expenditure in addition to that on capital assets. A five-year period is chosen to provide some smoothing of the effects of economic cycles and major capital expenditures.

Qualitative Factors. The recent history of a company should provide some insight into its future development. Past stability or growth in such key factors as Profit, ROCE, Margins, Sales, Productivity and EPS will be of particular interest. Although rates of growth can be expressed in numerical terms, judgement is needed in deciding what rates to expect in particular circumstances. These judgements apply both to past performance and to extrapolation into the future.

It must be recognised that some kinds of growth cannot go on forever. For example, a company is unlikely to sacrifice market share indefinitely in a prolonged attempt to maintain or improve its profit margin. Also, rising productivity following the introduction of more advanced equipment must be expected to peter out after a few years. Moreover, some irregularity in the growth of profits must be tolerated: since income is the difference between two much larger numbers, it is inevitably sensitive to changes in trading conditions.

Another relevant factor is the record of Extraordinary Items. The appearance of large write-offs occasioned by abandoning businesses may indicate habitually poor judgement by a company's board and managers.

Finally, indications of Contingent Liabilities and impending Legal Actions must be given serious consideration. Sometimes they are well known and have been allowed for by the creation of provisions and by a downward movement in the market price of a company's shares. Indeed, an investor may conclude that these corrections are excessive, and that investment is justified despite the company's looming problems.

Example 9.7 The General Electric Company plc 1991: Segment Order Books

---

In its 1991 annual report GEC included for the first time information on the orders to hand for each of its business sectors. The values given in Table 9.9 table provide insight into the health of each segment. The final column was not provided by the company, but has been calculated from the Turnover and Orders for each segment. It should be noted that each of the segments includes many businesses, some of which will have order books much longer than those for the whole segment, while for others trading is less buoyant.

As might be expected, the order books for capital goods and defence equipment are longest. Electronic Systems includes work on communications satellites and avionics. Power Systems includes railway vehicles, utility boilers, gas turbines and nuclear plant.

At the other end of the scale, Consumer Goods and Office Equipment display much leaner order books. This may arise from the fundamental nature of these businesses,

Table 9.10   Turnover and Order Book by Segment for Year Ending 31 March 91 (values in £ million)

	Turnover	Orders	Months of Sales
Electronic systems	2811	4720	20
Power systems	2530	4567	22
Telecommunications	1253	751	7
Consumer goods	271	14	½
Electronic metrology	449	64	1½
Office equipment and printing	339	19	½
Medical equipment	463	139	3½
Electronic components	346	169	6
Industrial apparatus	401	156	4½
Distribution and other	423	101	3
Totals	9286	10700	13.8

PLATE 13

TGV Atlantique train, built by GEC Alsthom and holder of the rail speed record.

which are dependent on the short time-scales on which merchandisers operate. However, it could reflect the recessionary conditions of 1990/91. When (as will presumably be the case) GEC has made available this kind of information for a number of years, it will be possible to construct time-series that will cast light upon these questions.

9.8 Nature of the Business: Growth or Cash?

Some of the parameters defined earlier provide insights into the fundamental nature of a company's activities. For example, the Value-Added Margin (Value-Added/Sales) is a measure of vertical integration. The average remuneration of employees is a measure of the quality of the work-force and thus of the sophistication of the company's products or services. It should be noted, however, that salaries and benefits are also strongly dependent on country of residence of the employees.

Technological Level. The level of a company's technological position and aspirations is measured by

Assets per employee
(R & D Spend)/Turnover
(R & D Spend)/Profit

In Example 9.6 Glaxo Holdings was found to display high values for each of value-added per employee, assets per employee, and average benefits per employee. Earlier, in Example 1.3, we noted the very high level of spending by this company on Research and Development. All of these indicators suggest that this is a company working at a high level of technology and determined to maintain that position.

It is the quality as well as the quantity of Research and Development that should interest the investor. One looks for evidence that a company's R & D is clearly directed towards the creation of improved products and production methods and hence to increased sales and profit margins.

Some companies maintain or improve their technological edge not by internal efforts, but by purchases of up-to-date equipment. Take as an example J Sainsbury, the food retailer. In recent years it has spent large sums on electronic point-of-sale equipment that speeds the check-out procedure and will provide improved stock control. Such purchases are the measure of that company's efforts to maintain its technological position.

Cash Generation. One sometimes sees a business referred to as a *Cash Cow*. The term is readily understood in a general sense: the activity generates stable sales and an acceptable profit margin, without requiring large capital reinvestment. Presumably the best measures of *Net Cash Generation* are parameters such as

Capital Requirement Ratio = (Capital Investment)/Retentions

or   = (Capital Investment + R&D Spend)/Retentions

Here Retentions means retained profit plus depreciation. Low values of such parameters characterise the cash cow, and high values a capital-absorbing business. One justification for the control of diversified businesses by a single holding company is the possibility of matching cash-generating and cash-absorbing activities.

In Section 9.7 other parameters bearing on cash generation were introduced.

Overseas Involvements. Another fundamental characteristic of a company is the extent of its international activity, which may take the form of overseas operations or overseas sales generated by exporting. A variety of ratios of components to total sales and of components to total profits can be used to measure these tendencies. The significance of these ratios depends upon a number of related factors: changes in exchange rates and relative inflation and growth rates, both for whole economies and for individual sectors.

A strong dependence on export sales in one country may be a sign of potential instability in profits. The example of Jaguar plc comes to mind. In the late 1980s a significant fraction of its sales of cars took place in the United States. The combination of a recession there and adverse movements in exchange rates subsequently depressed Jaguar's sales severely.

Substantial overseas trading can introduce difficulties in the interpretation of a group's accounts, particularly when inflation in regions of significant overseas trading is very different from that for the reporting currency. This is illustrated by the affairs of Polly Peck International plc, a holding company whose controlled activities extended from electronics (notably, the Japanese Sansui business) to trading in fresh fruit grown in northern Cyprus. After control of the company passed into the hands of administrators in 1990, it became apparent that inflation in Cyprus had distorted the accounts (or had allowed the accounts to be distorted), specifically by giving the impression that the company was highly profitable, while it was in fact absorbing more and more funds. Moreover, the administrators had difficulty in establishing control over cash the company was believed to have on deposit in the Middle East.

These unfortunate cases may give the impression that trading on an international scale inevitably increases the risk to which a company is exposed. This need not be so. Consider the Royal Dutch/Shell Group of companies. This integrated oil business is owned jointly by two holding companies, one based in Holland, the other in England. By virtue of the broad international spread of the trading of the group's subsidiaries, its profits are substantially insulated from currency fluctuations and economic misfortune in any one of its markets.

9.9 Investment Potential: To Buy or not to Buy?

Many of the parameters introduced in this chapter will help a potential investor decide whether to buy a company's shares or loan securities. Most were defined solely in terms of information contained in company accounts. The additional ratios to be introduced now involve also the market capitalisation of a company or the market price of its shares. As was explained in Section 7.3, it is the market return of Equation (7.11) that is ultimately of interest to the investor, and it is critically dependent on the share price.

Earnings-Related Parameters

Some ratios in general use and widely reported in the financial press are:

Earnings per Share (EPS)   (given in pence in the UK)

Price/Earnings (PER or P/E) = Market Price/EPS   (year)
= Market Capitalisation/Earnings   (year)

Earnings Yield = 1/PER   (1/year)

The units of these quantities are indicated in brackets to the right of the definitions. Earnings Yield was encountered earlier in Equation (7.9).

Earnings was defined in Section 4.2, as part of the consideration of the profit and loss statement, and EPS was introduced in Section 4.3. One sometimes encounters values of EPS that have been calculated from the *Maximum-Basis Earnings* or the *Nil-Basis Earnings* that were defined in Section 8.1. These values do not match exactly those determined using the *Net-Basis Earnings*, those found in the profit and loss statement.

EPS and PER values are usually based on so-called *Historic Earnings*, which are those most recently reported by the company. However, *Prospective Earnings* are sometimes used in the calculation; these are estimates of earnings in the present or later years.

The definition of these earnings-related measures of performance is somewhat complicated; their interpretation is even less straightforward. Suppose that, at the current market price, the PER for a company's shares is 13 (years). Such a value might be expected for a well-established engineering company in conditions of fairly buoyant trading. Taken at its face value, it suggests that the company will take 13 years to 'earn' the current price of its shares. The corresponding Earnings Yield is the reciprocal, 7.7 per cent (per annum). Viewed superficially, this may suggest that the purchase of the shares is an inferior investment to the alternative purchase of fixed-interest securities offering, say, 9 per cent (for government issues) or 10 per cent (for commercial issues).

However, interest received is subject to tax, which reduces its value to the recipient, while tax has already been deducted from reported earnings. Moreover, unless it is convertible or index-linked, a fixed-interest investment offers a specified repayment of capital; the return to redemption is thus strictly defined. On the other hand, over the medium to long term a stable company might be expected to grow in line with inflation, in sales and earnings and, most importantly, in respect of dividends and share price. These considerations show that the comparison of benefits of owning shares and loan stock must be approached more carefully.

A little algebra will help. What an investor presumably seeks through a particular share purchase is a set of circumstances that satisfies the inequality

$$EY + AG \geqslant 0.7\,(FI + RP) \quad (AT) \tag{9.1}$$

The factor 0.7 is introduced on the right-hand side to allow in a general way for the taxation applicable to typical investors. In practice, the incidence of capital-gains tax is variable, and some investors, notably charities, are not subject to income or corporation tax. As in Section 7.3, the label AT indicates after-tax returns, and BT before-tax returns.

On the left-hand side of Equation (9.1) is the *Expected Return*, comprising:
EY, the Earnings Yield, plus
AG, the anticipated rate of Additional Growth, beyond that generated by retaining part of the Earnings Yield.

The relationship (9.1) states that the sum is expected to equal or preferably to exceed the quantity on the right-hand side. The tax-reduced sum on the right comprises:
FI, the current rate of return on the most secure form of investment that is available, plus
RP, the Risk Premium demanded for equity investment in a particular company in the prevailing economic and market conditions.

Let us consider the earnings yield $EY = 7.7$ per cent in the light of this result, taking $FI = 9$ per cent. The inequality then becomes

$$AG \geqslant -1.4 + 0.7\,RP = 0.7\,(RP - 2)$$

This relates the assumed additional growth to the risk premium. In Section 7.3 we noted that a risk premium around 5 per cent (BT) represents the recent expectations of investors in the United Kingdom. (In other investment climates a different premium will be required.) We conclude that the typical investor expects the additional annual growth to be at least $AG = 2$ per cent (AT).

Looking at this argument in another way, we conclude that the investing community, in purchasing shares offering $EY = 7.7$ per cent, accepts a range of possible returns of which the expected return, some 9.8 per cent (AT), is the most probable. Corresponding to this expected return is an effective PER = 10.2 (years). However, given the risk inherent in this investment, the community of investors is simultaneously purchasing secure fixed-interest investments paying $FI = 9$ per cent (BT) or 6.3 per cent (AT).

It is interesting to determine also the corresponding real returns, by seeking to allow for inflation. For inflation rates of 4.5 to 5 per cent, levels typical of recent British experience, the real return on the equity investment considered above is around 5 per cent, while the real return from secure fixed interest is about 1.5 per cent.

To explore further the implications of the investment criterion of Equation (9.1), let us apply it to more highly rated shares, for which PER = 20 and $EY = 5$ per cent. We adopt a slightly higher risk premium of $RP = 6$ per cent, consistent with an anticipated higher volatility in the share price. Now Equation (9.1) gives

$$AG \geqslant 10.5 - 5 = 5.5 \text{ per cent (AT)}$$

Seemingly, the move to PER = 20 is equivalent to the expectation that additional growth $AG = 5.5$ per cent will be generated by the company. The anticipated return is now 10.5 per cent (AT), giving a pay-back period (or effective PER) of 9.5 years. Because the risk premium has been set a little higher, the anticipated pay-back period is a little less than that for the lower-rated shares considered above.

These examples demonstrate that it is misleading to interpret the raw Price/Earnings Ratio as the period over which the investor can expect to receive a cumulative return equal to the purchase price of shares.

Dividend-Related Parameters

The ratios of interest here, also reported regularly by the financial press, are:

Dividend per Share (usually given in pence in the UK)

$$\text{Dividend Yield} = \frac{\text{Ordinary Dividends (usually Gross)}}{\text{Market Price}}$$

$$\text{Dividend Cover} = \frac{\text{Earnings}}{\text{Ordinary Dividends (Net)}}$$

$$\text{Pay-Out Ratio} = \frac{1}{\text{Dividend Cover}}$$

The Dividend Yield is often referred to simply as the *Yield*. It is of great interest to the investor because it is actually remitted in cash, while much of the return represented by the Earnings Yield is (the investor hopes) turned into additional value within the company. The retained component can be realised only by selling the shares, and then only if their price has tracked the hoped-for performance of the company.

The *Dividend Cover* suggests how likely the company is to maintain or to increase its dividends. A high value could indicate that the company is likely to increase its cash pay-out to investors. But not necessarily. The directors may think it better to retain the funds within the company, arguing that an improvement in the share price will ultimately compensate investors at least as well as a large dividend. This is a specific example of a *Dividend Policy*. It expresses the view that the investor should not be particularly interested in the dividend, but should fix his attention on the total return.

In terms of the dividend and inflation rate, the investment criterion of Equation (9.1) can be expressed as

$$D + I + RG \geqslant 0.7\,(FI + RP) \quad (\text{AT}) \tag{9.2}$$

where D is the Net Dividend

I is the expected Inflation Rate, and

RG represents the Real Growth that the company is expected to achieve, beyond that required to match inflation.

This sum provides an alternative expression for the *Expected Return*.

In assuming a positive value for RG, an investor is hoping that a company's directors and staff will be able to create growth in excess of inflation by utilising the retained profit effectively, and by releasing additional value implicit in the company's operations and assets.

To demonstrate the implications, we consider again the specific cases discussed earlier. As before, we look at the case in which $EY = 7.7$ per cent and $RP = 5$ per cent, and now adopt the inflation rate $I = 5$ per cent. We take the Dividend Cover to be 2, so that $D = 7.7/2 = 3.85$ per cent. Introducing these values into Equation (9.2), we obtain

$$RG \geqslant 9.8 - 3.85 - 5 = 1 \text{ per cent}$$

Here about 40 per cent of the return is received immediately as dividends. The company is expected to do only a little better than keep up with inflation.

Turning to the second of the cases considered earlier, where $EY = 5$ per cent and $RP = 6$ per cent, we adopt a Dividend Cover of 3, so that $D = 5/3 = 1.7$ per cent. Then

$$RG \geqslant 10.5 - 1.7 - 5 = 3.8 \text{ per cent}$$

In this situation only 15 per cent of the anticipated return comes in the form of dividends. Internal growth is expected to be well in excess of inflation.

These calculations suggest much the same pattern of expectations as was revealed by Equation (9.1). The discussion has shown that investors commonly expect companies, having paid dividends, to more than match inflation through internally generated growth. Equation (9.2) indicates that the expectation of real growth is negligible when the dividend rises to 4.8 per cent (AT). For a pay-out ratio of ½, this corresponds to $EY = 9.6$ per cent and to PER = 10.4. Seemingly, a lower value of PER corresponds to the expectation of a reduction in the real value of the company and ultimately in the real value of its shares.

Maintaining constant dividend cover is not the only possible dividend policy. In the years following its privatisation British Gas plc adopted a *Progressive Dividend Policy*, with significant increases bringing down the cover year by year. Obviously, this process cannot continue indefinitely, but it does suggest that the directors of British Gas believed that a substantial fraction of the earnings should be remitted quickly to investors.

In hard times boards of directors seek to maintain the dividend, even if it is not covered by earnings. Experience has proved that investors do not quickly forgive a cut in the dividend. They are probably right to respond in this way. If a company cannot manage its affairs so that it can pay to investors this fraction of the expected return, one must wonder whether it can generate consistent internal growth.

The discussion based on Equations (9.1) and (9.2) will be reminiscent of that around Equation (7.6), which sought to define the cost of equity to a company. Here we have looked at the same problem, but from the point of view of the investor. In Section 7.3 the risk premium was characterised by the parameter Beta. The analysis above shows that measure of variability in returns to be related to the anticipated rate of growth in returns, measured by either *AG* or *RG*, different ways of expressing the ability of a company to generate profits.

Asset Backing

Another market-related parameter of interest to the investing community is the ratio

(Assets per Share)/Market Price

If he believes the published balance sheet, the investor can readily extract from it the value of Net Assets and determine this ratio directly. However, he may feel that the asset value should be adjusted upwards, to allow for failure to revalue or for unrecognised assets. Alternatively, he may have doubts about the possibility of realising the stated asset values in adverse circumstances.

By whatever means it is obtained, the Assets/Price ratio suggests whether the assets will provide a floor under the share price, and whether the investor is likely to get his money back if liquidation occurs. Investors are quite interested in these matters.

Another aspect of the balance sheet to be considered by the investor is gearing. Some analysts would argue that the ratio

Debt/(Market Capitalisation)

is a better measure of gearing than ratios based solely on balance-sheet values. We know that the reported equity is often much affected by accounting conventions. On the other hand, market capitalisation embodies the collective judgement of numerous independent observers whose minds are concentrated by the knowledge that their money is at stake. As we saw in Section 6.2, when considering British Aerospace and the BOC Group, this measure of gearing will be more stringent in some cases and easier in others.

Quality of Earnings

This phrase is encountered in discussions of the probability that the current level of profits can be maintained or improved upon. We can properly refer to the reported earnings as being of high quality if:

- They were struck after adequate allowance was made for depreciation, provisions, research and development, and maintenance of existing plant.
- They have not been inflated by a change in accounting conventions.
- Recent mergers or demergers do not obscure the accounts.
- The earnings display reasonable uniformity (preferably in growth) in the medium term, without evidence of undue massaging.
- There is no evidence of *Overtrading*, that is, of expansion so rapid that a cash shortage is likely, so that additional capital will be required.
- The earnings do not contain large overseas elements whose accessibility is doubtful.
- They are not excessively exposed to adverse exchange-rate risk.
- They appear to be supported by new products or by cost-reduction programmes.
- The company's operations are not exposed to an increasingly stringent regime of regulation by government-influenced agencies.
- Competition is not unusually severe, and the company appears well placed to counter such competition as exists.

If an investor demands assurance on every one of these points, the number of investments passing through the filter will be very small, and they may be of rather similar character. On the other hand, if more than one or two of these desired characteristics are absent, a cautious investor will feel happier with his funds placed elsewhere.

Example 9.8 Indicators Reported by Major Companies

All companies provide in their reports tables of key financial data for a five-year period. Some are merely values extracted from the basic financial statements for previous years. However, some retrospective presentations of performance go further, by providing a number of performance indicators of the kind we have been considering. The parameters chosen may be those which the company actually believes to be particularly significant. Alternatively, they may be ratios that the company thinks the reader will understand or will want to see. Whatever the reasons for providing this supplemental information, it is helpful to anyone wishing to understand where the company has come from and where it may be going.

A ratio that is always given is Earnings per Share, since it emerges directly from the profit and loss statement. The history of Dividends per Share is provided nearly as often; investors are obviously interested in a company's ability and willingness to give them money.

That is the extent of the common elements in the provision of information. Companies that offer more information typically provide two or three further indicators, but those chosen vary widely. The following values are found in about half of the reports that provide some analysis of past performance:

Trading (or Operating) Profit/Capital Employed
Trading (or Operating) Profit/Sales
Capital Gearing
Dividend Cover (that is, Dividends/Earnings)

Net Assets/Share
Interest Cover (that is, Interest/Profit BIT)
Capital Spending/Sales
Sales/Capital Employed
Return on SHF
Tax/Profit BT

No report provides all of this information. Ratios that are occasionally provided include:

It will be apparent that there is not a consensus among companies as to the selection of parameters that best characterise their performance and are therefore likely to be of greatest interest to readers of their reports.

10

What Next?

It is hard to know when to stop. Business finance is a diverse and rapidly evolving subject, with which engineers are much concerned now and will be more deeply involved in the future. The topics into which this book leads will be of crucial importance in the careers of many of its readers. However, in the space that remains we can do no more than note some of the loose ends left by this brief survey of finance and accountancy.

Chapter 6 says a good deal about ways of acquiring additional capital, but little about how it is to be used. The methods of Investment Appraisal weigh up the benefits and costs arising from investment in specific assets or projects. These techniques provide estimates of the returns to be expected from varied investments, and form a basis for the selection of one in preference to others. Investment appraisal also indicates whether the raising of additional capital is justified. If none of the suggested investments offers an acceptable return, there is no reason to seek more funds.

Another matter touched on in Chapter 6 is the interaction between companies and sources of capital, in particular, stock markets. The psychology of markets, and of those influenced by them, is a crucial feature of this relationship. At a more mundane level, the readily determined variability of market returns provides measures of the risk associated with individual companies, business sectors or entire economies. In fact, the world's economies, stock markets and investment pools are increasingly more closely linked, so that the returns from investments in different countries are moving closer together.

A major theme of Chapter 7 is the relationship between risk and borrowing, and in several places we have noted that risk is also influenced by diversification. A substantial body of Portfolio Theory has been developed to explore these relationships more fully. It can be argued that risk is reduced by moving into activities quite distinct from a group's existing business. However, such a diversification of interests is often achieved using borrowed funds. Is a diversified, but more highly geared company less exposed to risk than the original, more focused business? This is the kind of question addressed by portfolio theory. However, in finance as in engineering, it is wise to test theory against experience.

While considering the risk associated with business activity, one's thoughts turn to the most extreme form of risk – financial failure. A number of students of finance have addressed the problem of Failure Prediction, by analysing company accounts to see if they contain warnings of incipient collapse. Such warnings are unlikely to be explicit, for no company will willingly destroy the confidence of investors and customers by revealing its troubles with complete candour. The quest for a dependable predictor of corporate failure often involves a search for a magic combination of ratios selected from those introduced in Chapter 9.

Many of the examples of this book relate to changes in the ownership of companies, through privatisation, take-over, or a sale agreed between holding companies. Engineers find themselves bemused by abrupt changes in the ownership of the enterprises within which they work. These transfers of control seem sometimes to be generated by little more than political or corporate fashion. The programme of privatisation mounted in the United Kingdom during the 1980s is a giant experiment in corporate organisation. Its long-term consequences are unpredictable, but we can be sure that the future development of the privatised enterprises will provide insights into corporate life more generally. A study of the effects of re-nationalisation may also be required, and it could be equally instructive.

In Chapter 8 consideration was given to one of the most pervasive ways in which governments influence business activity, namely, through taxation. This affects companies not only by extracting a fraction of their profits, but through myriad steps they take to minimise the tax to be paid. We have already noted two gross forms of government interference with corporate life: nationalisation and privatisation. Less dramatically, governments – and increasingly the Commission of the European Community – see it as their duty to monitor and seek to control competition, in the belief that society's interests are best served by open and even-handed competition, extending across national boundaries.

The twentieth century has been marked by corporate Agglomeration, the formation of ever-larger groups, some being apparently arbitrary collections of disparate businesses. Evidently this process is not always advantageous, since these large groups often fail to achieve the business and financial aims that were expected when they were created. The future of some of the largest groups of engineering companies is under active discussion. If by some process, voluntary or imposed, GEC, BAe and ICI are broken down into smaller components, many of their employees will be profoundly affected.

Everything to do with Research and Development interests engineers, since these activities offer the most obvious opportunities to create something new. The funding and organisation of R&D are matters of continual debate, particularly in Britain, where research has traditionally flourished, while development, and design, have been perceived as weak. Comparisons of R&D expenditure in different companies, industrial sectors and economies may indicate why some sectors and companies flourish, while other decline. The nature and management of R&D activity are also matters warranting further study.

Most of the topics mentioned above come under the broad heading of Corporate Strategy. This gathers together the strands of finance, marketing, product creation and manufacture and addresses them in an international context. A consideration of corporate dynamics is the natural end point of the studies begun in this book. It is also the foundation of the healthy businesses within which engineering endeavours must be embedded if they are to succeed.

Further Reading

Accounting

Fanning, D. and Pendlebury, M., *Company Accounts – A Guide*, Allen and Unwin, 1984
Hitching, C. and Stone, D., *Understand Accounting*, Pitman, 1984
Parker, R. H., *Understanding Company Financial Statements*, Penguin, 3rd edn 1988
Rockley, L. E., *Business Accounting*, Heinemann, 1987

Companies

Bates, J. and Parkinson, J. R., *Business Economics*, Basil Blackwell, 1982
Chapman, K. and Walker, D., *Industrial Location – Principles and Policies*, Basil Blackwell, 1987
Hannah, L., *The Rise of the Corporate Economy*, Methuen, 2nd edn 1983
Hey, D. A. and Morris, D. J., *Industrial Economics – Theory and Evidence*, Oxford University Press, 1985
Johnson, P. (Ed.), *The Structure of British Industry*, Unwin Hyman, 2nd edn 1988
Lever, W. F., *Industrial Change in the United Kingdom*, Longman, 1987

Investment Analysis

Allen, D. H., *A Guide to the Economic Evaluation of Projects*, Institution of Chemical Engineers, 2nd edn 1980
Boyadjian, H. J. and Warren, J. F., *Risks – Reading Corporate Signals*, Wiley, 1984
Donnelly, G., *The Firm in Society*, Pitman, 2nd edn 1987
Rutterford, J., *Introduction to Stock Exchange Investment*, Macmillan, 1983

Periodicals

The Financial Times, daily
Business sections of *The Times*, *The Guardian*, *The Independent* and *The Daily Telegraph*, daily
The Economist, weekly
Professional Engineering, The Institution of Mechanical Engineers, monthly

Index to companies

Note   Page numbers set in **bold** type indicate examples relating to companies.

General index

Printed and bound by CPI Group (UK) Ltd, Croydon, CR0 4YY

01/11/2024

01782610-0013